International Conference on Environmental Engineering and Sustainable Development (CEESD 2016)

IOP Conference Series: Earth and Environmental Science
Volume 51

Sanya, China
9 – 11 December 2016

Editor:

Hsin Chu

ISBN: 978-1-5108-3595-5
ISSN: 1755-1307

Printed from e-media with permission by:

Curran Associates, Inc.
57 Morehouse Lane
Red Hook, NY 12571

Some format issues inherent in the e-media version may also appear in this print version.

Printed by Curran Associates, Inc. (2017)

For permission requests, please contact the Institute of Physics
at the address below.

Institute of Physics
Dirac House, Temple Back
Bristol BS1 6BE UK

Phone: 44 1 17 929 7481
Fax: 44 1 17 920 0979

techtracking@iop.org

Additional copies of this publication are available from:

Curran Associates, Inc.
57 Morehouse Lane
Red Hook, NY 12571 USA
Phone: 845-758-0400
Fax: 845-758-2633
Email: curran@proceedings.com
Web: www.proceedings.com

Table of contents

Volume 51

2016 International Conference on Environmental Engineering and Sustainable Development (CEESD 2016)

9–11 December 2016, Sanya, China

Accepted papers received: 10 January 2017
Published online: 4 February 2017

Preface

Environmental Engineering

Sustainable urban planning and development

Water and wastewater treatment

Hydrology

PREFACE

CEESD 2016 was held in Sanya, China during December 9-11, 2016. CEESD 2016 was organized by Asia Pacific Institute of Science and Engineering, sponsored by HKSME, supported by other organizations. The conference provides a useful and wide platform both for display the latest research and for exchange of research results and thoughts in broad Environmental Engineering and Sustainable Development. The participants of the conference were from almost every part of the world, with background of either academia or industry, even well-known enterprise. The success and prosperity of the conference is reflected high level of the papers received.

The proceedings are a compilation of the accepted papers and represent an interesting outcome of the conference. This book cover 4 chapters: Environmental Engineering; Sustainable urban planning and development; Water and wastewater treatment; hydrology.

We would like to acknowledge all of those who supported CEESD 2016. Each individual and institutional help were very important for the success of this conference. Especially we would like to thank the organizing committee for their valuable advices in the organization and helpful peer review of the papers.

We sincerely hope that CEESD 2016 will be a forum for excellent discussions that will put forward new ideas and promote collaborative researches. We are sure that the proceedings will serve as an important research source of references and the knowledge, which will lead to not only scientific and engineering progress but also other new products and processes.

Prof. Hsin Chu, National Cheng Kung University, Taiwan

COMMITTEES

Conference Committee Co-Chairs

Prof. Ruxu Du, The Chinese University of Hong Kong, Hong Kong

Prof. Irene Man Chi Lo, The Hong Kong University of Science and Technology, Hong Kong

Program Commitee Chairs

Prof. Chu Hsin, National Cheng Kung University, Taiwan

Prof. Zhi Chen, Concordia University, Canada

Prof. CHEN J. Paul, National University of Singapore (NUS), Singapore

Prof. Kansal, M.L.Indian Institute of Technology - Roorkee, India

Publication Chairs

Prof. MARK GIBSON, Dalhousie University, Canada

Prof. Yang fan, Hainan University, China

International Technical Committee

Prof. Kit Ming LAM, the University of Hong Kong, Hong Kong (Chair)

Prof. Rajasekhar Balasubramanian, National University of Singapore (NUS), Singapore

Prof. Wai Yuen SZETO, the University of Hong Kong, Hong Kong

Prof. Sumit Sen, Indian Institute of Technology – Roorkee, India

Dr. Tsang, C.W. Daniel, The Hong Kong Polytechnic University, Hong Kong

Dr. Sayed M. Bateni, University of Hawaii at Manoa, USA

Dr. Nicky Y. F. LAM, City University of Hong Kong, Hong Kong

Dr. IRVAN DAHLAN, University of Science Malaysia, Malaysia

Dr. Huang mingzhi, Sun Yat-sen University, China

Peer review statement

All papers published in this volume of *IOP Conference Series: Earth and Environmental Science* have been peer reviewed through processes administered by the proceedings Editors. Reviews were conducted by expert referees to the professional and scientific standards expected of a proceedings journal published by IOP Publishing.

Characterization, Degradation, and Reaction Pathways of Indoor Toluene over Visible-light-driven S, Zn Co-doped TiO_2

H Chu[1,*], Y H Lin[1,2] and C Y Lin[1]

[1]Department of Environmental Engineering, National Cheng Kung University, Tainan 70101, Taiwan
[2]College of Ecology and Resource Engineering, Wuyi University, Fujian Wuyishan 354300, China

* E-mail: chuhsin@mail.ncku.edu.tw

Abstract. Sulfur and Zinc co-doped TiO_2 prepared by a sol-gel method to degrade toluene under a fluorescent lamp was investigated. The results indicate that S,Zn co-doped TiO_2 photocatalysts are mainly nano-size with an anatase phase structure. The degradation reactions of toluene were performed under various operation conditions. The results show that the toluene conversion increases with increasing toluene concentration and decreasing relative humidity. Based on the results of activity test, $S_{0.05}Zn_{0.001}/TiO_2$ was chosen for further studies. The main oxidation products of toluene photodegradation are CO_2, H_2O, benzyl alcohol, acetone, butadiene and acetic acid. Two possible mechanisms have been developed for photodegradation of toluene in a dry and a humid environment.

1. Introduction

Indoor air pollution has received increasing attention in recent years since people spend almost 90 percent of their time indoors [1]. Volatile organic compounds, which are usually emitted from construction materials, household products, and office furnishings, are the main indoor air pollutants [2]. Toluene, as one of the most predominant VOCs, is widely applied to rubber, paint, ceiling, disinfectant, tackiness agent, and chemical reactants. Toluene has high toxicity, carcinogenicity and long persistence in the indoor environment. Therefore, the removal of toluene is an important task for the indoor air quality. Photocatalytic oxidation (PCO) using TiO_2 has recently received considerable attention for VOCs removal due to its cost-effectiveness, non-toxicity, fast oxidation rate, and chemical stability [3]. This study was conducted under various operating conditions. The characteristics of photocatalysts, photodegradation of toluene, and pathway of toluene photooxidation have been investigated under a fluorescent lamp.

2. Materials and methods

2.1. Photocatalyst preparation

Sulfur and Zinc co-doped TiO_2 with various mole ratios were prepared by a sol-gel method. In the preparation process of Ti-precursor sol, titanium tetra-isopropoxide (TTIP) (30 mL) was mixed with absolute ethanol (45 mL). Then the mixture was slowly added into deionized water (360 mL) at room temperature during continuous stirring. After mixture was stirred vigorously at room temperature for 30 min, 1.2 ml of HNO_3 (65 wt%) was slowly added dropwise into the mixture, which was further stirred until the gel formed. The final solution was heated on a hot plate set at 70°C and constantly stirred until the liquid fully evaporated. Finally, the residual was calcined at 500°C for 3 h in air. On the basis of the previous study [3], the S/Ti molar ratio was chosen as 0.05 and the Zn/Ti molar ratio was y. In the preparation of $S_{0.05}Zn_y/TiO_2$, the pre-determined amounts of 0.37 g thiourea (CH_4N_2S) and the pre-determined amounts of zinc nitrate ($Zn(NO_3)_2{\cdot}4H_2O$) was dissolved in 360 mL de-ionized water first, and the subsequent processes were the same as the method described above.

2.2. Photocatalyst characterization

The XRD patterns were obtained on a Rigaku X-ray diffraction (Model D/MAS IIIV, Rigaku) using Cu Kα radiation to determine the major crystal phases. The crystallite sizes of the photo-catalysts were calculated by using Scherrer's formula [4].

2.3. Photocatalytic activity measurements

The reaction gas of toluene was generated using a programmable syringe pump. The reaction gas of toluene was mixed with cylinder air and water vapor according to target toluene concentrations, ranged from 0.5 to 2 ppm. The reaction gases before and after the experiments were analyzed with a gas chromatography (GC)/flame ionization detector and fitted with a DB-624 capillary column. A 10W fluorescent lamp (FL-10D, emission range between 360 and 580nm with λ_{max} = 436 nm) was used as the photon source and the light intensity was 2.5 mW cm^{-2}. The by-products and products were analyzed using an on-line FTIR spectrophotometer (Model Spectrum One, Perkin Elmer), operating in the spectral range of 4,000 to 700 cm^{-1}.

3. Results and discussion

3.1. XRD analysis

The crystal phases, average crystalline sizes, and lattice parameters of various photocatalysts are listed in Table 1. The crystal phase of bare TiO_2 and co-doped TiO_2 are all anatase phases checked from JCPDS Card No. 21-1272. The crystallite size of bare TiO_2 is about 16 nm, and those of TiO_2 co-doped with sulfur or zinc are about 11 nm. The doping reduces the grain size, which is attributed to the lattice distortion caused by the incorporation of S and Zn into the TiO_2 matrix. The phenomena can be demonstrated from the change in lattice parameter of TiO_2. The results show the occurrence of a slight lattice distortion in the crystal structure. For $S_{0.05}/TiO_2$, the ionic radius of S^{6+} (0.12 Å) is smaller compared to that of Ti^{4+} (0.68 Å), which can substitute Ti^{4+} ions in the TiO_2 crystal lattice, and leads to the formation of Ti–O–S bond. The S,Z co-doped TiO_2, Zn^{2+} ions do not enter into crystal lattice of TiO_2 because ionic radius of Zn^{2+} (0.74 Å) is larger than Ti^{4+} (0.68 Å) in the TiO_2 structure [5]. Therefore, Zn^{2+} ions are dispersed on the surface of the S/TiO_2 particles in the form of ZnO clusters during the calcination process.

Table 1. Crystallite structure characteristics of various samples.

Sample	Crystal phase Anatase (%)	Crystallite size d_A (nm)	Lattice parameter a (Å)	c (Å)	c/a
TiO_2	100	16.6	3.777	9.517	2.520
$S_{0.05}/TiO_2$	100	11.9	3.783	9.492	2.509
$S_{0.05}Zn_{0.0001}/TiO_2$	100	11.7	3.787	9.483	2.504
$S_{0.05}Zn_{0.001}/TiO_2$	100	11.8	3.792	9.504	2.507
$S_{0.05}Zn_{0.01}/TiO_2$	100	11.1	3.793	9.471	2.497

3.2. Photocatalytic degradation of toluene

Photocatalytic degradation of gaseous toluene was investigated in a continuous-flow reactor. Figure 1 shows the removal efficiencies of toluene by various photocatalysts (operating conditions: $[O_2]$ = 21%, [toluene] = 2 ppm, [RH] = 60%, residence time = 30 sec, Temp. =35°C). The degradation results show that co-doped TiO_2 have better photocatalytic activities compared to S-doped TiO_2 and bare TiO_2. Among them, $S_{0.05}Zn_{0.0001}/TiO_2$ exhibits the highest photocatalytic activity. The higher degradation rates of toluene by S,Zn co-doped and S-doped TiO_2 may be caused by the substitution of Ti atoms by sulfur atoms to form Ti–O–S bonds in TiO_2. The formation of Ti–O–S possibly leads to band gap narrowing, and then increases the absorption of visible light irritation. However, lower photocatalytic activities of photocatalysts with zinc contents above 0.0001% are obtained. The phenomena is ascribed to that active sites of photocatalyst surface are occupied by excess zinc. Therefore, $S_{0.05}Zn_{0.0001}/TiO_2$ was chosen to further study its photocatalytic reaction.

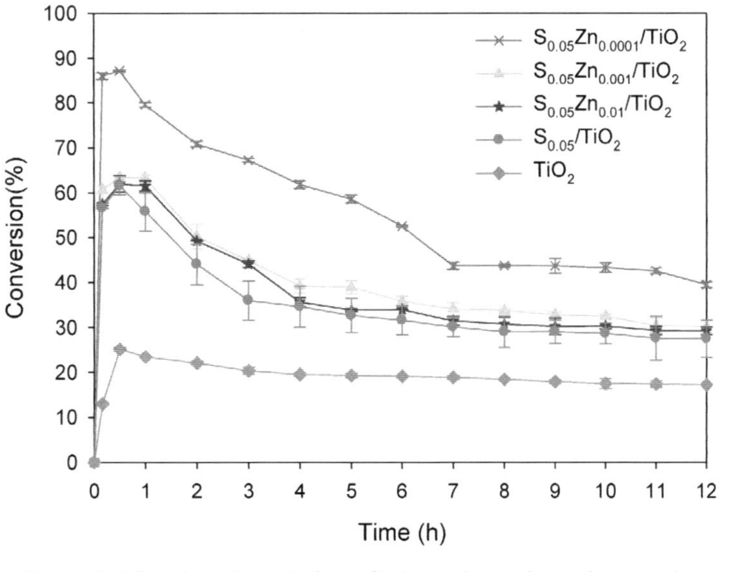

Figure 1. The photodegradation of toluene by various photocatalysts.

3.3. Gaseous products and byproducts by FTIR

In order to investigate byproducts and products of toluene photodegradation, a series of FTIR observations have been performed and the results are shown in Figures 2. IR spectra were demonstrated with reference from the database (Linstrom and Mallard, 2001). The relative humidity were set at two levels (RH = 0, 60% with $[toluene]_{in}$ = 2 ppm). The absorption peaks around 3,500–4,000 and 1,350–1,900 cm^{-1} are attributed to H_2O, which is formed during the photocatalytic degradation of toluene or water vapor originally in the gas stream. The peaks intensity of water presents higher strength in a wet condition than in a dry condition. The absorption spectra located at 3,067 and 3,100 cm^{-1} are the characteristic peaks of butadiene. The absorption bands assigned at about 2,955, 3,011, and 1,270 cm^{-1} are attributed to acetone. The band at about 2,360 is assigned to gaseous CO_2. The toluene is observed at about 1,100 cm^{-1}. The broad absorbance in the range between 740 and 870 cm^{-1} can be assigned to benzyl alcohol. The absorption bands at about 640–660 cm^{-1} are characterizing of acetic acid.

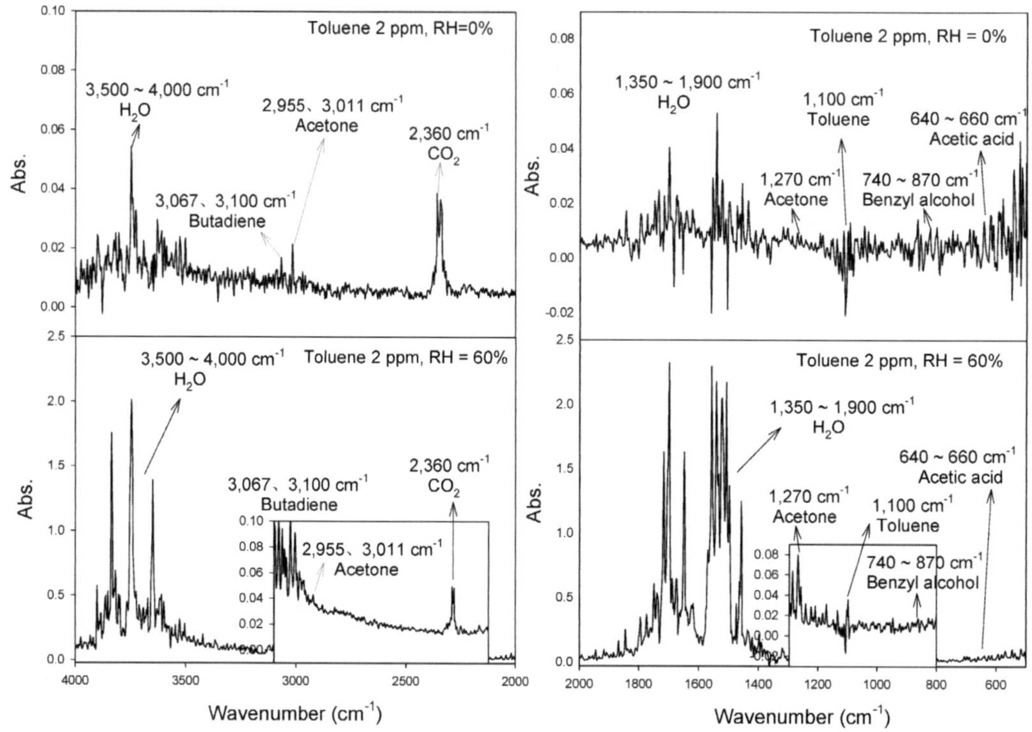

Figure 2. FTIR spectra of byproducts and products of toluene photodegradation

3.4. Reaction pathway

The decomposition of toluene occurs by attacking from superoxide or hydroxyl radicals, and its selectivity strongly depends on the reaction conditions. Superoxide anion radicals and electrons ($\bullet O_2^-/e^-$) play the key roles for the rate limiting step of the reaction in low level humidity. In a high level humidity, hydroxyl radicals and holes ($\bullet OH/h^+$) are the key roles. Figure 3 illustrates two potential reaction pathways in dry and wet condition. In the case of dry condition, the production of photo-induced electrons and holes is achieved when a lamp is turned on. The negatively charged electrons react with adsorbed oxygen molecules to form superoxide radicals. At the beginning of the experiment, benzyl radical ($\bullet C_7H_7$) may be formed through $\bullet O_2^-$ radical attack [6]. The $\bullet O_2^-$ radical attack also leads to the formation of a long chain volatile organic compound. A further attack by $\bullet O_2^-/e^-$, those byproducts can generate benzylperoxy radical, benzyl alcohol, benzaldehyde, benzoic acid, and long-chain organic compounds. The $\bullet O_2^-/e^-$ attack also results in the formation of Acetone or butadiene. Acetone and butadiene are also oxidized by $\bullet O_2^-$ to form acetic acid, carbon dioxide, and water. Hydroxyl radicals are considered the key species for the wet reaction process. The positively charged holes generate hydroxyl radicals ($\bullet OH$) by reacting with adsorbed water, and then initiate the oxidation of the adsorbed organic compounds. The reaction pathway of the wet condition is similar to that for a dry condition except that it may generate cresols no production of benzylperoxy radical and aromatic bridged peroxo intermediate.

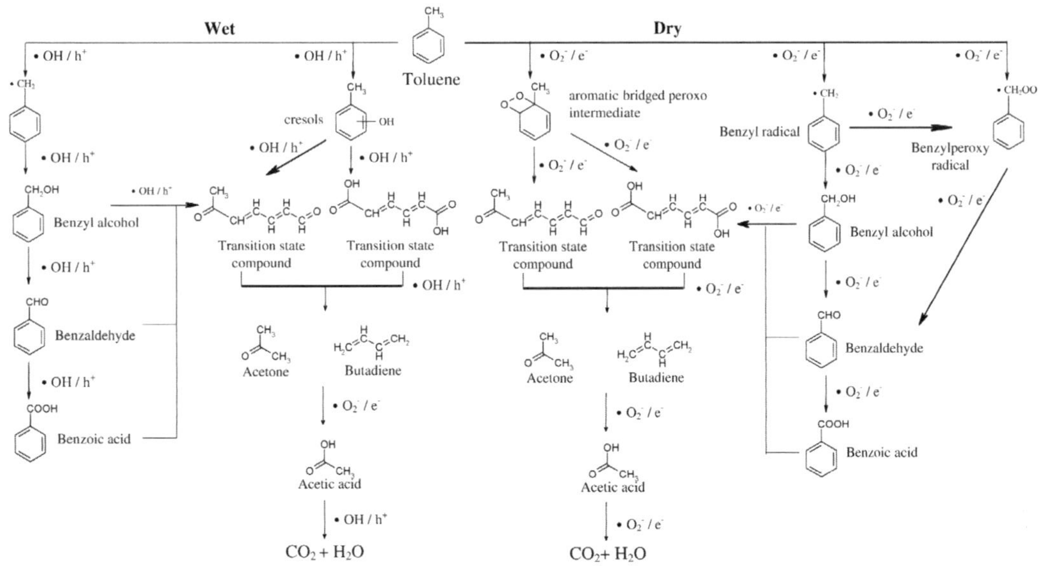

Figure 3. Proposed reaction pathways for photocatalytic conversion of toluene in dry and wet condition

4. Conclusion

The results show that S,Zn co-doped TiO_2 can reduce the crystalline size of the photocatalysts. $S_{0.05}Zn_{0.0001}/TiO_2$ exhibits the best decomposition rate for toluene. The decomposition rate of toluene decreases with increasing relative humidity and decreasing temperature. The products of the toluene photocatalytic oxidation are CO_2 and water; the by-products include benzaldehyde, benzyl alcohol, benzoic acid, acetone, butadiene and acetic acid. Superoxide radicals and •OH radical are considered the key species for the dry and wet reaction process, respectively.

Acknowledgments

This study was funded in part by the Ministry of Science and Technology, Republic of China, under grant NSC 104-3113-E-006-004

References

[1] Spengler J D and Sexton K 1983 *Science* **221** 9-17.
[2] Puddu V, Choi H, Dionysiou D D and Puma G L 2009 *Appl. Catal. B: Environ.* **94** 211-8.
[3] Lin Y H., Tseng T K and Chu H, 2014 *Appl. Catal. A: Gen.* **469** 221-8.
[4] Ogawa H, Abe A, Nishikawa M and Hayakawa S 1981 *J. Electrochem. Soc.* **128** 685-9.
[5] Yang L, Zhang Y, Ruan W, Zhao B, Xu W and Lombardi J R 2010 *J. Raman Spectrosc.* **41** 721-6.
[6] Sleiman M, Conchon P, Ferronato C and Chovelon J M 2009 *Appl. Catal. B: Environ.* **86** 159-65.

Numerical Investigation of Nanofluid Laminar Forced Convective Heat Transfer inside an Equilateral Triangular Tube

Amin Etminan[1,*], **Zambri Harun**[2] **and Ahmad Sharifian**[3]

[1, 2] Department of Mechanical and Materials Engineering, Faculty of Engineering and Built Environment, 43600 UKM Bangi, Selangor, Malaysia
[3] School of Mechanical and Electrical Engineering, University of Southern Queensland, Toowoomba QLD 4350, Australia

[*]E-mail: etminan.amin@gmail.com

Abstract. In this article distilled water and CuO particles with volume fraction of 1%, 2% and 4% are studied numerically. The steady state flow regime is considered laminar with Reynolds number of 100 and nanoparticles diameters (dp) are set in the range of 20 nm and 80 nm. The hydraulic diameter and the length of equilateral triangular channel are 8 mm and 1000 mm respectively. The problem is solved using finite volume method with constant heat flux for two sides and constant temperature for one side. Convective heat transfer coefficient, Nusselt number and convective heat transfer coefficient distribution on walls are investigated in details. The fluid flow is supposed to be one phase flow. It can be observed that nanofluid leads to a remarkable enhancement on heat transfer coefficient pressure loss through the channel. The computations reveal that the size of nanoparticles has no significant influence on heat transfer properties. Besides, the study shows a good agreement between current results and experimental data in the literatures.

1. Introduction

The thermal industrial equipment has experienced unprecedented improvements in enhancing heat transfer capabilities and this has resulted in new challenges for thermal sciences especially with the advancement of nanotechnology. Nanotechnology as well as nanoparticles have been used in a wide variety of industries. In recent years, nanofluids have become popular in heat exchanger installations for achieving higher heat transfer rates. Nanoparticles are suspended in a base fluid and it is called nanofluid [1]. Nanofluids are completely different from microfluids and they have distinctive features in comparison to conventional solid-liquid mixtures in which small sized particles in the range of mm or μm of metals and non-metals are dispersed [1]. Knowledge about nanofluids using in heat transfer enhancement mechanism is still in its primary. A new class of polymer nanofluids, drag-reducing nanofluids, aim at enhanced heat transfer as well as flow friction reduction have been studied by Phelan et al [2] where a wide range of active self-assembly mechanisms for nanoscale structures has been investigated in details.

Khaled et al [3] studied the laminar heat transfer regime in a channel with and without nanoparticles. They found an obvious enhancement for the case where nanoparticles have been added to the base fluid. Heris et al [4] investigated a numerical method for laminar heat transfer in a channel with constant heat flux on all walls. Their results clearly showed that addition of nanoparticles to the base fluid produces remarkable enhancement of heat transfer. Nanoparticle concentration can increase heat transfer coefficients and decreasing their size can enhance heat transfer coefficients as well. Wen et al [5] focused on the study of Al_2O_3-water nanofluid in the laminar flow regime with constant heat flux on walls of a coppery pipe. Their results showed that the increase in Reynolds number and nanoparticle size with 1.6% volume fraction caused 47% enhancement of Nusselt number. Several different numerical methods including Euilerean, one-phase and two-phase-mix methods were used by Lotfi et al [6] for exploring the effect of a vast variety of volume fractions of nanoparticles on heat transfer parameters. Ahmad et al [7] studied heat transfer in three nanofluids which nanoparticles Al_2O_3, CuO and SiO_2 were suspended in the base fluid of ethylene glycol with nanoparticles concentrations ranged from 1 to 6% in a channel with triangular cross section numerically. Flow regime was considered laminar with Reynolds number of 100-800. Their results disclosed 50% rise of Nusselt number when Reynolds number changed from 100 to 800. In an experimental research carried out by Kim et al [8], the heat transfer of both laminar and turbulent nanofluid flow regimes in a channel with constant heat flux on walls was studied. They found that Al_2O_3-water nanofluid with volume fraction of 3%, caused 8% rise in conduction and 20% increase in convective heat transfer coefficients.

In summary, the heat transfer of none-circular cross section channels is generally less than that of circular ones [9]. In the present work, the finite volume method has been used for simulation of water and CuO nanofluid with volume fraction of 0%, 1%, 2% and 4%. The steady state laminar flow regime with Reynolds number of 100 and nanoparticles diameters of 20 nm and 80 nm are considered. The convective and diffusive terms in Navier-Stokes equations have been simulated by a second order upwind method in all simulations meanwhile the SIMPLE procedure has been employed for the velocity and pressure relation. The fully developed x-velocity and uniform temperature T_0=300 K is assumed at the tube inlet meanwhile the fully developed conditions are assumed at the tube outlet. The stationary wall conditions and uniform heat flux are imposed on the solid walls of tubes. The geometry of equilateral triangular cross section tube with hydraulic diameter of 8 mm and length of 1000 mm can be observed in Figure 1.

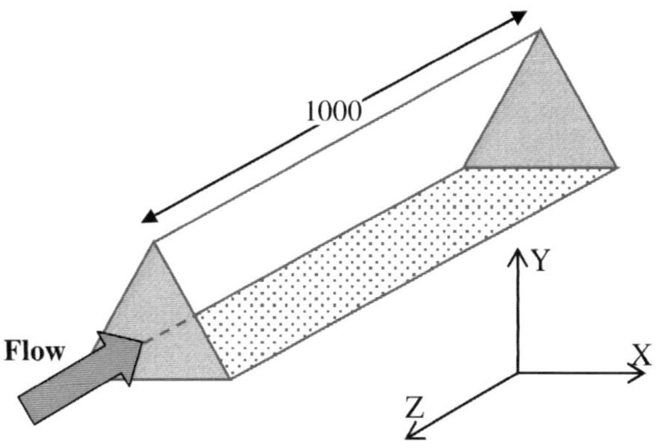

Figure 1. Geometry of problem

2. Governing Equations and Numerical Simulation Approach

The governing equations in single phase model are similar to those of the base liquid equations. The effective conservation equations including continuity, momentum and energy for steady state flow regime are as follow (All following symbols, subscripts and Greek letters are tabled in nomenclature section at the end of the paper.):

$$\nabla.(\rho v) = 0 \tag{1}$$

$$\nabla.(\rho v v) = -\nabla p + \nabla.(\mu \nabla v) \quad (2)$$

$$\nabla.(\rho c_p v T) = \nabla.(k \nabla T) \quad (3)$$

where thermo physical properties including μ, ρ, k, c_p in before equations should be replaced with effective values μ_{eff}, ρ_{eff} , k_{eff} , $c_{p,eff}$ respectively which are listed as below [10,11]:

$$\mu_{eff} = (123\,\emptyset^2 + 7.3\,\emptyset + 1)\mu_f \quad (4)$$

$$\rho_{eff} = (1-\emptyset)\rho_f + \emptyset\rho_p \quad (5)$$

$$k_{eff} = \{1 + 64.7\,\emptyset^{0.7460}\left(\frac{d_f}{d_p}\right)^{0.3690}\left(\frac{k_s}{k_f}\right)^{0.7476}\mathrm{Pr}^{0.9955}\mathrm{Re}^{1.2321}\}k_f \quad (6)$$

$$c_{p,eff} = (1-\emptyset)c_{p,f} + \emptyset c_{p,p} \quad (7)$$

Prandtl and Reynolds numbers are described as $\mathrm{Pr} = \frac{\mu}{\rho_f \alpha_f}$ and $\mathrm{Re} = \frac{\rho_f \beta_c T}{3\pi\mu^2 L_{bf}}$ respectively.

In above equation, L_{bf}=0.17 nm and β_c=1.3807E-23 are the mean free path of water and Boltzmann constant for entire tested temperature range of water respectively. Furthermore, thermal diffusivity and dynamic viscosity are calculated by $\alpha_f = k_f/\rho_f c_{p,f}$ and $\mu = a10^{b/(T-c)}$, $a = 2.414 \times 10^{-5}$, b=*247* and c=*140*, [11].

Diameter of water molecule is calculated by $d_f = 0.1\left(\frac{6M}{N\pi\rho_f}\right)^{1/3}$ where $M = 18\ gr/mol$

Thermo physical properties of water and nanoparticles are listed in Table 1.

Table 1. Thermo physical properties of water and nanoparticles.

Thermo physical properties	Water	CuO
ρ (kg/m^3)	997	6300
C_p (J/kgK)	4181.7	537
μ (Pa.s)	0.00089	-
k (W/mK)	0.6069	17.65

Constant heat flux and constant temperature of 320 K are considered on two sides of channel and horizontal side respectively. The average convective heat transfer coefficient and Nusselt number for constant temperature plate are defined as follow:

$$\bar{h} = \frac{\dot{m}c_p(T_{m,o} - T_{m,i})}{A_s\,\Delta T_{lm}} \quad (8)$$

$$\overline{NU} = \frac{\bar{h}D_h}{k_{eff}} \quad (9)$$

where ΔT_{lm} means LMTD and it is defined by the below relation:

$$\Delta T_{lm} = \frac{(T_s - T_{m,o}) - (T_s - T_{m,i})}{\ln[(T_s - T_{m,o})/(T_s - T_{m,i})]} \quad (10)$$

To find a proper as well as independent grid for solving the problem of heat transfer in a three dimensional channel, several different grid distributions have been evaluated and the average temperature of outlet water is computed for three different grid sizes as in Table 2.

Table 2 Comparison of average water temperature in outlet flow for different grid sizes

Grid size(nm)	Node numbers	T (K)
1	14 × 12 × 1000	306.012
0.8	17 × 15 × 1000	305.543
0.6	23 × 20 × 1000	305.490

It can be seen that the average temperature of water in outlet flow for the grid size of 1 nm has not adequate accuracy, however, the grid size of 0.8 nm and 0.6 nm are much accurate and are very close to each other. Therefore, the grid size of 0.8 nm is chosen in order to have less calculation time.

3. Numerical Results

Forced convective heat transfer in a channel with equilateral triangular cross section and CuO-water nanofluid for several volume fractions and two nanoparticle sizes of 20 nm and 80 nm has been investigated extensively. With introducing a heat flux ratio that shows heat flux relation between two sides of channel ($r_q=q_2/q_1$). The horizontal side is considered with constant temperature of 320 K. As it can be seen in Figures 2 and 3 convective heat transfer coefficient and average Nusselt number versus heat flux ratio for different volume fractions are grown linearly. Furthermore, the rise of heat flux ratio leads to growth of average convective heat transfer coefficient and average Nusselt number (See Figures 2 and 3). This work is done for three different heat flux ratio of 0, 0.5 and 1. The Reynolds number for all cases is equal to Re=100 and average convective heat transfer coefficient and average Nusselt number are obtained for horizontal side of the tube.

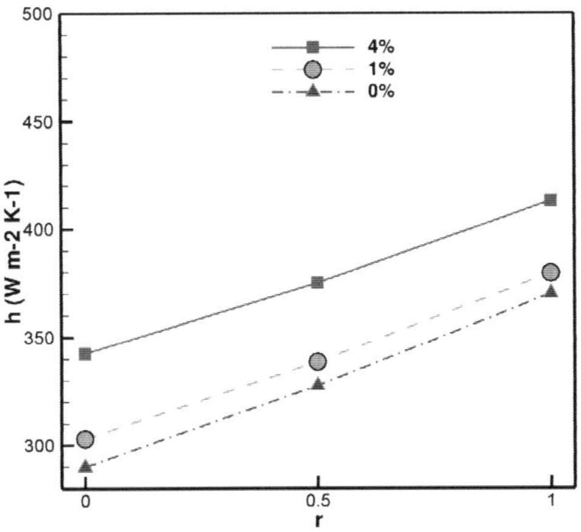

(a)

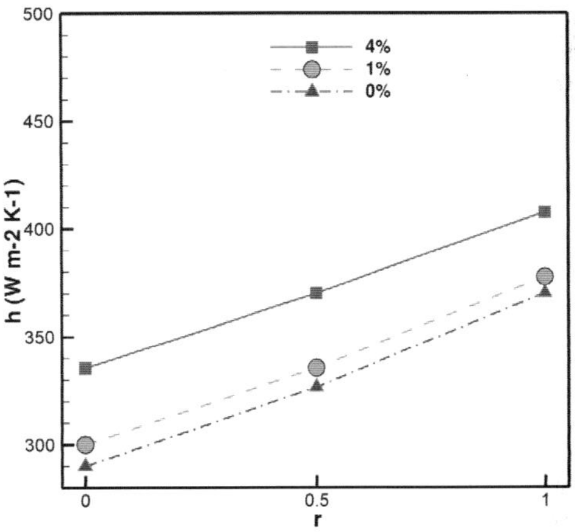

(b)

Figure 2. Effect of heat flux ratio on average convective heat transfer coefficient for diameter of nanoparticles (a) 20 nm and (b) 80 nm

According to Figures 2 and 3, it can be found that the average convective heat transfer coefficient increases versus varying of nanoparticle diameter from 80 nm to 20 nm. For instance, average convective heat transfer coefficient of nanofluid with 4% volume fraction, heat flux ratio of 1 and nanoparticle size of 20 nm and 80 nm rise 11.6% and 10% respectively in comparison with distilled water. Furthermore, average convective heat transfer coefficient of nanofluid with 4% volume fraction, heat flux ratio of 0 and nanoparticle size of 20 nm and 80 nm increase 18% and 16% respectively in comparison with the distilled water. A decline in nanoparticle size from 80 nm to 20 nm diminishes average Nusselt number. For example, average Nusselt number of nanofluid with 4% volume fraction, heat flux ratio of 1 and nanoparticle size of 20 nm and 80 nm records 1.64% and 4.1% increases respectively in comparison to the distilled water. As an another example average Nusselt number of nanofluid with 4% volume fraction, heat flux ratio of 0 and nanoparticle size of 20 nm and 80 nm show 7.8% and 9.69% growths respectively in comparison to water. Totally decreasing nanoparticle size leads to increasing average convective heat transfer coefficient and decreasing average Nusselt number. So decreasing nanoparticle size does not lead to much more heat transfer. For a constant nanoparticle size, by decreasing heat flux ratio from 1 to 0, average Nusselt number increases. So unbalancing heat fluxes on walls leads to a better convective heat transfer and Nusselt number as well.

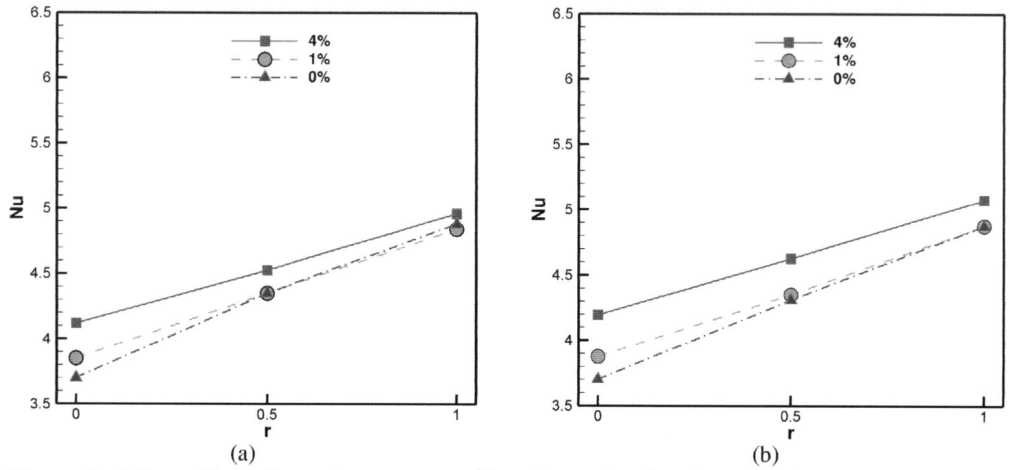

Figure 3. Effect of heat flux ratio on average Nusselt number for diameter of nanoparticles (a) 20 nm and (b) 80 nm

On the other hand, Figures 4 (a) and (b) show local convective heat transfer coefficient and Nusselt number on hot plate for heat flux ratio of 1 and nanoparticle size of 80 nm along the channel respectively. The plotted local convective heat transfer coefficient in Figure 4(a) for three volume fractions of nanofluid depicts the much more differences at channel inlet in comparison to that of outlet and becomes less with moving along the channel.

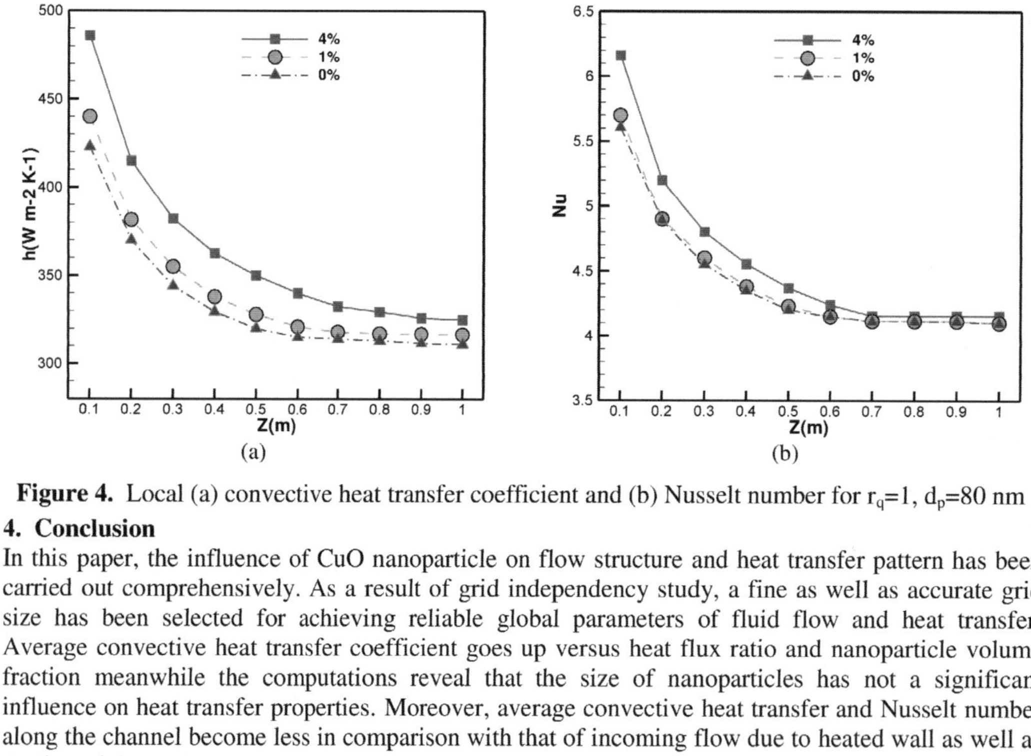

Figure 4. Local (a) convective heat transfer coefficient and (b) Nusselt number for $r_q=1$, $d_p=80$ nm

4. Conclusion

In this paper, the influence of CuO nanoparticle on flow structure and heat transfer pattern has been carried out comprehensively. As a result of grid independency study, a fine as well as accurate grid size has been selected for achieving reliable global parameters of fluid flow and heat transfer. Average convective heat transfer coefficient goes up versus heat flux ratio and nanoparticle volume fraction meanwhile the computations reveal that the size of nanoparticles has not a significant influence on heat transfer properties. Moreover, average convective heat transfer and Nusselt number along the channel become less in comparison with that of incoming flow due to heated wall as well as decline of thermal boundary layer. Furthermore, in the presence of nanoparticles, a significant rise on heat transfer coefficient and pressure loss through the channel is observed where the maximum magnitude of velocity is computed at the centre of channel cross section.

Acknowledgment
This research is financially supported by Fundamental Research Grant Scheme, Ministry of Education, Malaysia.

Nomenclature

C_p	Specific heat [J/kgK]
d	Diameter [m]
d_p	Nanoparticle diameter [m]
D_h	Hydraulic diameter of the triangular tube [m]
h	Convective heat transfer coefficient [W/m^{2}K]
k	Conduction heat transfer coefficient [W/mK]
L	Length [m]
N	Avogadro's number [$= 6.022 \times 10^{23}$]
NU	Nusselt number
p	Pressure [Pa]
Pr	Prandtl number
q''	Heat flux [W/m^2]
Re	Reynolds number
T	Temperature [K]
T_m	Fluid mean temperature [K]
T_s	Solid wall temperature [K]
r_q	Heat flux ratio
x	Cartesian coordinate axis [m]
y	Cartesian coordinate axis [m]
z	Cartesian coordinate axis [m]
Greek Letters	
α	Thermal diffusivity [m^2/s]
μ	Dynamic viscosity [kg/ms]
ν	Viscosity [m^2/s]
ρ	Density [kg/m^3]
$\emptyset$	Volume fraction [%]
Subscript	
ave	Average
bf	Mean free path
eff	Effective
f	Fluid
i	Inlet flow
lm	Logarithm mean temperature difference
m	Mean
o	Outlet flow
p	Nanoparticle
s	Solid wall

References

[1] Singh A K 2008 *Thermal Conductivity of Nanofluids* Defence Science Journal, Vol. 58, No. 5, pp 600-607.
[2] Phelan P E, Bhattacharya P, Prasher R S 2005 *Nanofluids for heat transfer applications* Annual Review of Heat Transfer, Vol. 14, pp 255-75.
[3] Khaled A R A and Vafai K 2005 *Heat transfer enhancement through control of thermal dispersion effects* International Journal of Heat Mass Transfer, Vol. 48, No. 11, pp 2172-2185.
[4] Heris S Z, Esfahany M N and Etemad G 2007 *Numerical Investigation of Nanofluid Laminar Convective Heat Transfer through a Circular Tube* Numerical Heat Transfer, Part A: Application, Vol. 52, No. 11, pp 1043-1058.
[5] Wen D and Ding Y 2004 *Experimental Investigation into Convective Heat Transfer of Nanofluids at the Entrance Region under Laminar Flow Conditions* International Journal of Heat and Mass Transfer, Vol. 47, No. 24, pp 5181-5188.
[6] Lotfi R, Saboohi Y and Rashidi A M 2010 *Numerical study of forced convective heat transfer of Nanofluids: comparison of different approaches* International Communication in Heat and Mass Transfer, Vol. 37, No.1, pp 74-78.
[7] Ahmed H E, Mohammed H A and Yusoff M Z 2012 *Heat transfer enhancement of laminar Nanofluids flow in a triangular duct using vortex generator* Superlattices and Microstructures, Vol. 52, No.3, pp 398-415.
[8] Kim D, Kwon Y, Cho Y, Li C, Cheong S, Hwang Y, Lee J, Hong D and Moon S 2009 *Convective Heat Transfer Characteristics of Nanofluids under Laminar and Turbulent Flow Conditions* Current Applied Physics, Vol. 9, No. 2, pp 119-123.
[9] Aberoumand H, Jahani A, Aberoumand S and Jafarimoghaddam A 2015 *Experimental Investigation on Heat Transfer of Silver-Oil Nanofluid in Concentric Annular Tube* Transport Phenomena in Nano and Micro Scales, Vol. 3, No. 2, pp 118-128.
[10] Zabihi K, Gholamian F and Vasefi S I 2013 *Experimental and Numerical Investigation of Al_2O_3-Water Nanofluid Inside a Triangular Tube* World Applied Sciences Journal, Vol. 22, No. 5, pp 601-607.
[11] Drew D A and S. L. Passman S L 1999 *Theory of Multicomponent Fluids* Springer.

Research Progress of Artificial Forest in the Remediation of Heavy Metal Contaminated Soils

MA Jiafang[1], MENG Guangtao[2], HE Liping[2]* and LI Guixiang[2]

[1] College of environmental science and engineering, Kunming University of Science and Technology, Kunming Yunnan 650093
[2] Yunnan Academy of forestry, Kunming Yunnan 650204,

* E-mail: kmhlp@sina.com

Abstract. (1) Remediation of soil contaminated by heavy metals has become a hot topic in the world, and phytoremediation technology is the most widely used. (2) In addition to traditional economic benefits, ecological benefits of artificial forest have been more and more important, which are very helpful to soil polluted with heavy metals in the environment. (3) The characteristics of heavy metal pollution of soil and plantations of repair mechanism have been reviewed, and the current mining areas, wetlands, urban plantations on heavy metal elements have enriched the research results. The purpose is to find a new path for governance of heavy metal soil pollution.

1. Introduction

With the social development, a large number of metal materials are used. At the same time, serious pollution has also occurred. Heavy metal pollution has become a serious environmental problem. Many researchers are focused on a series of control technologies. At present, the best response is the phytoremediation technology, which is a biological treatment method. This technology features low cost and small disturbance to the environment. It has important significance in the treatment of heavy metal pollution of soil. Many plants have been studied as repair materials, including herbs, shrubs and woody plants. Much attention has been paid to research reports regarding the remediation of many woody plants contaminated by heavy metals, gradually mature woody repair mechanism, and artificial forest as a special community of woody plants. So this paper reviews the current research progress in remediation of heavy metal pollution on artificial forest soils, to provide more references for future related research.

2. Heavy Metal Pollution of Soil and Remediation Technologies

These heavy metals, including Cu, Zn, Pb, Mn, Cd, Hg, Ni, As, would cause soil pollution. According to reports, in 1950 there were2.2 million tons of cadmium (Cd) emitted to air globally, 93.9 million tons of copper (Cu), 78.3 million tons of plumbum (Pb) and 1.35 million tons of zinc (Zn)[1]. For example, Cu^{2+} is a necessary element for plants[2], but excessive Cu^{2+} would cause serious harm to plants, thus leading to intracellular biological metabolism, and ultimately inhibiting plant growth[3].

Soil suffers the most serious heavy metal contamination. These metals eventually enter into the soil through ecological cycles, causing heavy metal pollution of soil to different degrees. For example, according to preliminary statistics, China currently has 26.67 million hm^{-2} farmland subjecting to different degrees of heavy metal pollution, and the contaminative farmland accounts for 20% of the total farmland; Pb, Cd, Hg, Cu and combined pollution are most prominent[4]. Meanwhile, crops and other plants are grown in contaminated soil through the food chain cycle. Ultimately heavy metal elements in crops and other plants will enter the body, and eventually pose hazards to human body because of its biodegradable, highly toxic, significantly cumulative effect. So, prevention of heavy metal pollution of soil is extremely important.

Now remediation techniques are mainly around the physical, chemical, biological methods. To repair soil contaminated with heavy metals, physical methods are commonly used. New soil is used to replace old soil. Later, a heat-treatment is used for electric repair. The chemical treatment method is mainly based on soil chemical properties. It adds soil improvers, surfactants, inhibitors and other chemical reagents, in order to reduce the activity of heavy metals. The bioremediation method uses biological metabolic activity to reduce the concentration from harmful substances in the environment[5]. To treat heavy metal pollution, the physical and chemical methods were mainly adopted in the past, but physical and chemical methods are complex, costly, and likely to cause secondary pollution, so bioremediation is at the forefront of research. The most commonly used biological method is phytoremediation, which uses plants to absorb heavy metals. It is efficient and economical. In recent years, as the repair material plant has been used more and more, we find that the use of heavy metals and Hyperaccumulators tolerant plants is the key to phytoremediation.

3. Repair of artificial forest polluted by heavy metals

3.1 Restoration mechanism of artificial forest

The artificial forest is a special forest resource. Compared to natural forests, it has some advantages, such as fast growth and ease to use. Initially, planting artificial forest aims to meet the social demand for timber products, but due to ecological construction, artificial forest not only brings economic benefits, but also has more application areas, such as large quantities having been planted on the mining area, coastal wetlands, urban green land, desert, greenbelt and protected zones.

Many hyperaccumulators are herbaceous, but with phytoremediation technology development and discovery of hyperaccumulators[6], research shifted focus to woody plants[7]. So, in recent years, artificial forest has begun to play the role of phytoremediation for soils polluted with heavy metals. Artificial forest is woody plants, especially trees, which have large biomass, roots, stems, leaves, and dry areas having greater exposure to the environment. It can absorb, transform and degrade pollutants, and does not easily spread pollutants to the environment. Most artificial forest mainly aims at producing timber and the green environment. Even if the trees have enriched some heavy metal elements, it will not easily enter the food chain and cause harm to human health. Because of this reason, woody repair industry has emerged in many countries[8]. In China, over the past ten years, more and more studies showed that many trees play a special role in repairing heavy metal pollution.

3.2 Heavy metal pollution repair of the mining artificial forest

Mineral exploitation, especially some metal mining areas, can cause heavy metal pollution problems. Many studies have been focused on vegetation restoration mine. From 1999 the arsenic (As) hyperaccumulators[9], Sedum alfredii Hance and zinc (Zn) hyperaccumulators[10] were found in China; a variety of hyperaccumulators were planted, and artificial forest has been widely used in many mining areas.

In Malanzhuang iron ore of Tangshan, such measures as engineering reclamation and biological reclamation are adopted, which have initially established the artificial forest ecosystems[11]. It has not only improved the ecological environment in mining and restored vegetation, but also absorbed heavy metals iron mine (Fe) to one degree or another. In manganese mine wastelands of Xiangtan planted artificial *Koelreuteria paniculata*[12], and analyses show that different ages of *Koelreuteria paniculata* can absorb Mn, Fe, Zn, Pb, Cu; and these heavy metals in Koelreuteria paniculata showed decomposition coefficients greater than migration coefficient. In zinc (Zn) ore area, *Leucaena leucocephala (Lam)*[13] not only survived, but also enriched 80% zinc in the xylem, which make heavy metal elements fail to migrate to other environments, thus avoiding the surrounding environment from being affected by pollution, and causing no plantations treatment of heavy metals in the mining area.

3.3 Heavy metal pollution repair of artificial mangroves

Mangroves are important coastal vegetation. It has a series of ecological roles, such as cleaning sea and air, protecting banks, dissipating wave, and being habitats and feeding grounds for many birds[14]. So, the coastal wetland system plays an important role. But now as the coastal tidal wetlands are increasingly polluted by heavy metals, the relationship of mangrove ecosystems and heavy metal pollutants has been studied, and these studies show that mangroves have some resistance and tolerance of various pollutants, and the mangrove forest has the ability to absorb pollutants and clean water[15]. By measuring three species of mangrove plants (*Aegiceras corniculata, Kandelia candel (Linn) Druce, Acanthusilicifoliu*) from Maowei Sea Mangrove Wetland of Guangxi[16], it has been found that the adsorption capacities of metal elements (including: Cu, Zn, Cr, Ni, Pb, CD, Hg and As) of three plants are different, *Aegiceras corniculata* : Zn>Cr>As>Pb>Cu>Ni>Cd>Hg; Kandelia candel(Linn)Druce:Zn>Cr>Cu>As>Pb>Ni>Cd>Hg;Acanthusilicifoliu:Zn>As>Cr>Cu>Pb>Ni>Cd>Hg. These studies show that mangroves play an important role of controlling heavy metals. Especially in China's northern part (Fujian Province) there is *Kandelia candel (Lin.) Druce* ecosystem[17]. By analyzing its heavy metals and pollutant distribution, the artificial forest has great heavy metal adsorption capacities. In 2014, some scholars analyzed seven mangrove plants and their root heavy metal sediments (Cu, Zn, Cd, Pb, As), and the results show that in addition to weak transport capacities, other heavy metals have strong transport capacity. The article recommends introducing strong enrichment plants[18].

3.4 Heavy metal pollution repair of urban artificial forest

Suitable artificial forest is planted in urban areas, which not only play the role of greenbelt, but also prevent heavy metal pollution. There are research data showing that *Pinus massoniana* Lamb has a 23 years of planting history in urban areas in Guangxi[19]. These artificial trees totally accumulate six heavy metal elements, totaling 39.791kg/hm^2, where Mn, Zn, Pb, Cu, Cd, Ni elements reserves were 34.047, 3.351, 1.226, 0.874, 0.245, 0.084 kg/hm^2, reserves of heavy metals in the components of the spatial distribution of dry> bark> root> leaf> branch, indicating that *Pinus massoniana Lamb* could accumulate heavy metals, especially Mn. The same study also comes from the analysis of the accumulation and migration of *Cinnamomum camphora (L) Presl* forest of Hunan. Different heavy metal elements have different accumulation capabilities, and the general trend is Cd> Mn> Zn> Cu> Pb> Ni, and dry, bark, root, leaf, branch have different accumulation capabilities: dry>bark>root>branch>leaf. In other words, there are some heavy metals accumulated in the not easy moving parts of artificial forest. In densely populated areas, the human body could contact with heavy metals. Forests planted in many urban areas accumulated heavy metals from air and soil to trunk, and such forest clean the environment and avoid heavy metal elements from entering the food chain.

4. Conclusion

Compared with natural forests, the research applications of artificial forest have some undesirable factors, such as short planting years and unstable forest ecosystem. But now, with the progress of research and construction of artificial forest, the role of artificial forest has been continuously excavated in environmental governance.

So the problem should be studied more in the future:

(1) The plantation ecological benefits in terms of preventing acid rain, and other organic pollutants.

(2) Further screen artificial forest to find more hyperaccumulators.

(3) Technology of repairing heavy metal pollution is characterized by low investment, sustainability and ecological safety, which conform to the concept of ecological construction, but for the artificial forest that accumulates heavy metals, direction to use should be considered after harvest.

(4) Woody repair technology should be combined with other techniques, such as woody - herbaceous united phytoremediation, and physico-chemical methods.

Acknowledgements:

This study was supported by the NSFC (Natural Science Foundation of China) "Mechanism of soil quality variation in the process of vegetation restoration area of phosphorite Yunnan mining wasteland"(41361076).

References:

[1] Xin W and Qixing Z 2004 The ecological process, effect and remediation of heavy metals contaminated soil *Ecological Science* **23(3):** 278-81

[2] Yijun W, Xiaoyun L, Jian W 2014 Study on the Metabolism Pathway of ROS and NO of Tomato Seedlings Mediated by Exogenous NO under Copper Ion Stress Asian *Journal of Ecotoxicology* **9(4):**678-88

[3] Wojcik M and Tukiendorf A 2003 Response of wild type of Arabidopsis thaliana to cooper stress *Biologic Plantarum* **46:**79-83

[4] Dongbai W, Jian Z, Ping w Yan L, Wenhao X 2014 A Review in situ Immobilization of Heavy Metals Contaminated Soil Using Environmental Materia *Chinese Agricultural Science Bulletin* **30(8):**181-85

[5] Wiles S C 1993 Bioremediation of soil contaminated with polynucler aromatic hydrocarbons *Environmental pollution* **81**:229-49.

[6] Liang X, Ming G, Shengqing S 2014 Physiological Response and Accumulation Characters of Four Anti-pollution Woody Plants under Antimony Stress *Ecology and Environment Sciences* **23(8):** 1344-50.

[7] Xiuwen W, Bin Y, Huiwen L 2011 Application of Dendroremediation to the Soil Contaminated Soil by Heavy Metals and Organic Pollutants *Forest Research.* **47(5):** 124-30.

[8] Sell J, Kayser A, Schulin R 2005 Contribution of Ectomycorrhizal Fungi to cadmium Uptake of Poplars and Willows from Heavily Polluted Soil *Plant and Soil.* **277(12) :** 245-53

[9] Strycharz S and Newma L 2009 Use of Native Plants for Remediation of Trichloroethylene: II. Deciduous Trees *International Journal of Phytoremediation.* **11(2):**150-70.

[10] Xinxian L, Xiao Y, Zhengqian Y 2002 Differences of Uptake and Accumulation of Zinc in Four Species of Sedum. *Acta Botanica Sinica.* **44(2):** 152-57.

[11] Fuping L,Dong X,Tingzhong L 2010 Research on Ecological Rehabilitation of Dump in Malanzhuang Iron Mine. *Metal Mine.* **2**:152-54.

[12] Zhaohui L, Dalun T, Hongdeng T 2013 Biological Cycling of Koelreuteria paniculata Plantation Microelements in Xiangtan Manganese Mine Wasteland. *Acta Ecologica Sinica.* **20:** 6517-25.

[13] Zhingquan Z, Minghong H, Xiangping N 2000 The Revegetation of A Lead / Zinc Mine Tailings Site with An Introduced Soil Seed Bank. *Chinese Journal of Plant Ecology.* **24(5)**:97-104.

[14] Weidong H, Xiumei G, Changyi L 2000 The Ecological Values of Mangrove Ecosystems in China. *Ecological Science.* **19(1):**40-45.

[15] Xiaoju L, Yuanxiao J, Guizhu C 2005 Advances on Pollution Ecology and Purification Effect of Mangrove Wetland System. *Wetland Science.* **3(4):**315-19.

[16] Junjie G, Liangliang H, Zhiqiang W 2015 Heavy Metal Accumulation Characteristics of Wetland Plants in the Mangrove of Maowei Gulf. *Journal of Guilin University of Technology.* **35(1):**138-41.

[17] Dongjin H, Weibin Y, Ren W 2013 Spatial Distribution of Heavy Metals in *Kandelia candel* Wetland Sediments of Natural Forest and Plantation in the North Marginal Region of Chinese Natural Mangroves. *Chinese Journal of Applied & Environmental Biology.* **19(6):**945-51.

[18] Peng W, Zhingzhong Z, Ronglin M 2014 Bioaccumulation Characteristics of Heavy Metal in Intertidal Zone Sediments from Northern Hainan Island. *Ecology and Environment Sciences.* **23(5):**842-46.

[19] Xi F, Dalun T, Wenhua X 2004 Absorption,accumulation and dynamic of heavy metal elements in Pinus massoniana plantation in Guangxi. *Guihaia.* **24(5)**:437-42.

Review on Nano zerovalent Iron (nZVI): From Modification to Environmental Applications

Xing Chen, Dandan Ji, Xiqing Wang and Lihua Zang*

College of Environmental Science and Engineering, Qilu University of Technology, University Town, Changqing District, Ji'nan City, Shandong Province, China

*E-mail: zlh20151216@126.com

Abstract. Nanoscale zero-valent iron (nZVI), with its reductive potentials and wide availability, offers degradative remediation of environmental contaminants. Although nZVI particles were successfully used in wastewater treatment, there are still some drawbacks associated with the process and needs to be addressed. For example, magnetic attraction between nanoiron particles causes the rapid aggregation of particles. In addition, nZVI are more prone to react with dissolved oxygen and oxygen-rich compounds. Meantimes, nZVI could exert some degree of toxicity towards microbial species, and the effects of nZVI at the cellular and community levels are progressively being elucidated. To overcome these problems, nZVI particles were immobilized in or on suitable solid supports, and also to expand the effective pH range of the Fenton reaction. Therefore, refer to the new trends is very important when selecting nZVI for the treatment of various matrices .This review also identifies problems that may occurs a result of changes in the physicochemical properties of nZVI due to their modification (e.g. other metal doping, coating the surface, or deposition on the support) . Toxicity studies suggest that cellmembrane disruption and oxidative stress through the generation of Fe^{2+} and reactive oxygen species by nZVI are the main mechanisms contributing to nZVI cytotoxicity. This review highlights the application of nano-zero valent iron in treating refractory compounds.

1. Introduction

The multidisciplinary nano 'boom' has led to the evolution of a wide array of novel technologies for both domestic and industrial applications, including improved drug delivery and new methods for the treatment of contaminated water.[1] As the particle size decreases, proportion of atoms at the surface increases, raising its tendency to adsorb, interact, and react with other atoms, molecules, and complexes to achieve charge stabilization. The potential use of nanoparticles for the treatment of contaminated groundwater has sparked a great deal of interest.

Nanoscale zero valent iron (nZVI) is the most commonly used nanomaterial in Europe and in the United States for soil and groundwater remediation. Due to its reduced size, nZVI has a higher reactivity towards a broad range of contaminants, including halogenated compounds, nitrate, phosphate, polycyclic aromatic hydrocarbons, and heavy metals, and a higher mobility compared to its microscale counterpart. Consequently, nZVI is regarded as a promising remediation strategy suitable to a broad range of applications and environments.

O'Hara et al. and Quinn et al. reported substantial reductions of trichloroethylene (TCE) in soil (greater than 80% reduction) and groundwater (60% to 100% reduction) during a fieldscale demonstration at Cape Canaveral Air Force Station (Florida), by injecting emulsified nZVI particles.

In view of the abundant published literature on the topic, this review provides an updated overview of the potential impacts of the in situ deployment of nZVI on microbial communities (summarized in Figure 1).

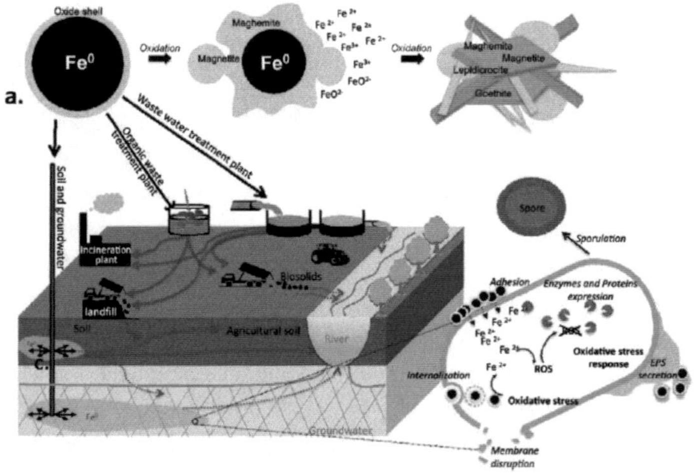

Figure 1. nZVI oxidation, cytotoxicity, cellular defense mechanisms mediated by nZVI, and potential routes of nZVI in the natural environment. Illustration of (a.) nZVI oxidation process, (b.) bacterial toxicity and defense mechanisms, and (c.) potential routes followed by nZVI in the environment. Plain gray arrows represent deliberate injection or amendment of nZVI, and deliberate transport of material potentially containing nZVI. Dashed gray arrows represent potential non-deliberate transport of nZVI in non-target environments. Abbreviations: EPS, Extracellular polymeric substances; ROS, Reactive oxygen species.

The key properties essential for the use of any engineered nanoparticle for in situ remediation includes (a) high reactivity for contaminant removal, (b) sufficient mobility within porous media, (c) sufficient reactive longevity, and (d) low toxicity. However, it should be noted that the process involved is at a sensible cost and competitive with other existing technology.[2, 3]

The practical applicability of ZVI lies in the fact that it can easily get oxidized to +2 and +3 oxidation states and in the process reduce other organic as well as inorganic impurities. Metallic iron easily acts as an electron donor:

$$Fe^0 \rightarrow Fe^{2+} + 2e^- \quad (1)$$

Nanoscale iron particles are in a process to replace micro iron particles and have proven to be quite effective reductant and catalyst for a wide variety of common environmental contaminants including chlorinated organic compounds and metal ions.[4]

2. Application of nZVI for remediation of organic compounds

Use of nZVI particles in advanced oxidation processes (AOPs) such as the Fenton Process have shown an advantage over conventional method which requires around 40–80 ppm of ferrous ion in the solution and this value is above the standards. In addition, the application of homogeneous AOPs to treat large quantity of water may produce large amount of sludge in the final neutralization step.[5]

In order to avoid these disadvantages, nano-zero valent iron (nZVI) could be used as an alternative way to induce Fenton oxidation.

2.1. Fenton like reactions using nano-zero valent iron

In heterogeneous Fenton reaction, oxidation of nZVI provides an alternative means of inducing Fenton oxidation[6, 7] as shown in Eq. (2):

$$O_2 + Fe^0 + 2H^+ \rightarrow Fe^{2+} + H_2O_2 \quad (2)$$

nZVI is the reactive reagent, and applications in vadose-zone soil or shallow groundwater rapidly oxidize nZVI to ferrous (Fe^{2+}), as shown below, and ferric iron (Fe^{3+}) leading the released electron to become available for the reduction of other compounds, whereas, through Fenton chemistry (Eqs. (2) and (3)), which produces strong oxidants capable of transforming recalcitrant environmental contaminants such as chlorinated compounds, pesticides, polychlorinated biphenyls, inorganic anions and heavy metals.[8, 9]

$$Fe^{2+} + H_2O_2 \rightarrow Fe^{3+} + OH^- + \cdot OH \quad (3)$$

Use of nZVI to induce Fenton oxidation has some advantages beyond addition of Fe^{2+}/H_2O_2[10]:

a.nZVI is able to attach or coat on large particle, therefore nZVI absorbed media could treat contaminated water passing through a sand filter or other type of filtration system.

b.nZVI injected through wells could be immobilized in/on soil grains in contaminated aquifers.

c. an increase in the reductive degradation reaction rate.

d. a decrease of the reductant dosage.

e. control over the risk of release of toxic intermediates.

f. the generation of a nontoxic end product.

Joo et al [6] studied about the ability of nZVI to induce oxidative degradation of contaminants. This study demonstrated that nZVI was capable of degrading the herbicide molinate in the presence of oxygen. In this study, it was shown that nZVI-induced Fenton oxidation could degrade over 60% of molinate at pH 8.1 and 65% at pH 4.

Bergendahl and Thies [10] have investigated the feasibility of oxidative removal of methyl-tert-butyl ether (MTBE) using ZVI. The results showed that over 99% of MTBE in water was degradedwithin 10 min and the major oxidation product was acetone. The addition of H_2O_2 greatly enhanced the degradation efficiency. The rate of degradation of MTBE increases with an increase in H_2O_2:MTBE ratio.

Although nZVI particles were successfully used in wastewater treatment, there are still some drawbacks associated with the process and needs to be addressed. For example, magnetic attraction between nanoiron particles causes the rapid aggregation of particles. In addition, nZVI are more prone to react with dissolved oxygen and oxygen-rich compounds. To overcome these problems, nZVI particles were immobilized in or on suitable solid supports, and also to expand the effective pH range of the Fenton reaction. Now widely used load materials is PVA microsphere[11, 12]. The others include calcium alginate beads [13], iron exchanged with Nafion membranes[14], iron modified clays[15, 16]. The modified supporting materials such as silica fabric[17], zeolites[18], resins [19], and cotton[20] which are used for immobilization of iron ions could be used for immobilizing nZVI.

2.2. nZVI cytotoxicity

Modifications of nZVI increased stability, reactivity, mobility, and reduced aggregation or passivation of nZVI. However, modifications of nZVI may lead to the creation of materials that not only eliminate contaminants more effectively, but also accumulate in living organisms, migrate over large distances sedimentation on bottoms of water reservoirs, or become carriers of other contaminants. Studies have demonstrated that the toxicity effects of nZVI are limited compared to other nanoparticles.[21]

The toxicity of nanoparticles can be caused by a variety of factors, largely convergent with those which are responsible for their high reactivity. The available information on nZVI is focused mainly on their effect on microorganisms and to a lesser extent, on crustaceans, fish larvae, arthropods, annelids, plants, and occasionally mammalian cells. Studies show that nZVI can be adsorbed on cell membranes of bacteria, or penetrate through them, which often leads to disturbances in the functioning of the cell. [22] Nanoparticles adsorbed on cell membranes can block cellular ducts, cause structural changes to the membranes, or inhibit mobility and nutrient intake and result in death of the bacteria. Research shows that modifications of the nZVI surface can be improved in their stabilization, reduction of aggregation, and reduction of toxicity. All forms of nZVI aggregated in soil and water in presence of a high concentration of calcium ions and thus addition of calcium salts may be improved in reducing the toxicity of groundwater due to nZVI .[23]

Another often-suggested mechanism contributing to nZVI toxicity involves the generation of highly reactive oxygen species, which accumulate in the cell environment and denature macromolecules including lipids, proteins, and nucleic acids, damaging intracellular structures and eventually leading to cell death. Because under aerated conditions, Fe^{2+} oxidizes more rapidly than it would under anaerobic conditions, the contribution of Fe^{2+} to nZVI toxicity is higher under anaerobic conditions than under aerobic conditions.[24] Modification of nZVI, e.g. as a result of nZVI coating with polyasparaginate, reduces the toxicity of nZVI by limiting direct contact of nanoparticles with cells.

3. Conclusion

The investigation of nZVI particles on organic compounds degradation is receiving more attention in the current decade. As research demonstrates, nZVI can be an effective and versatile tool for the purification of waters and soils. nZVI was successfully used for the generation of hydroxyl radicals in

AOP system. However, the potential threats resulting from the application of modified and nonmodified nZVI to various matrices are rather poorly recognized. Another important aspect in the estimation of the safety of application of nZVI is to determine how the various methods of modification will affect the fate of nZVI and their effect on living organisms.

Acknowledgement
This work was supported by the National Key Technology Research and Development Program of the Ministry of Science and Technology of China (Grant No. 2014BAC25B01, Grant No. 2014BAC28B01), and the Technology Development Plan of Jiangsu Province, China (Grant No. BE2015662).

References

[1] BYSTRZEJEWSKA-PIOTROWSKA G, GOLIMOWSKI J, URBAN P L. 2009 Nanoparticles: Their potential toxicity, waste and environmental management. *Waste Management*, 29(9): 2587-95.
[2] ZHANG W X. 2003 Nanoscale Iron Particles for Environmental Remediation: An Overview. *Journal of Nanoparticle Research*, 5(3-4): 323-32.
[3] CRANE R A, SCOTT T B. 2011 Nanoscale zero-valent iron: Future prospects for an emerging water treatment technology. *Journal of hazardous materials*, 211-212(211-212): 112-25.
[4] LIEN H L, ZHANG W X. 2001 Nanoscale iron particles for complete reduction of chlorinated ethenes. *Colloids & Surfaces A Physicochemical & Engineering Aspects*, 191(1): 97-105.
[5] IURASCU B, SIMINICEANU I, VIONE D, et al. 2009 Phenol degradation in water through a heterogeneous photo-Fenton process catalyzed by Fe-treated laponite. *Water research*, 43(5): 1313-22.
[6] JOO S H, FEITZ A J, WAITE T D. 2004 Oxidative degradation of the carbothioate herbicide, molinate, using nanoscale zero-valent iron. *Environmental Science & Technology*, 38(7): 2242-7.
[7] BABUPONNUSAMI A, MUTHUKUMAR K. 2012 Removal of phenol by heterogenous photo electro Fenton-like process using nano-zero valent iron. *Separation & Purification Technology*, , 98(98): 130-5.
[8] GIASUDDIN A B M, KANEL S R, CHOI H. 2007 Adsorption of humic acid onto nanoscale zerovalent iron and its effect on arsenic removal. *Environmental Science & Technology*, , 41(6): 2022-7.
[9] KEENAN C R, SEDLAK D L. 2008 Factors affecting the yield of oxidants from the reaction of nanoparticulate zero-valent iron and oxygen. *Environmental Science & Technology*, 42(4): 1262-7.
[10] BERGENDAHL J A, THIES T P. 2004 Fenton's oxidation of MTBE with zero-valent iron. *Water research*, 38(2): 327-34.
[11] XUE B, YE Z F, QU Y Z, et al. 2009 Immobilization of nanoscale Fe 0 in and on PVA microspheres for nitrobenzene reduction. *Journal of hazardous materials*, 172(2-3): 1357-64.
[12] BABUPONNUSAMI A, MUTHUKUMAR K. 2012 Treatment of phenol-containing

wastewater by photoelectro-Fenton method using supported nanoscale zero-valent iron. *Environmental Science & Pollution Research*, , 20(3): 1596-605.

[13] BEZBARUAH A N, KRAJANGPAN S, CHISHOLM B J, et al. Entrapment of iron nanoparticles in calcium alginate beads for groundwater remediation applications. *Journal of hazardous materials*, 2009, 166(2–3): 1339-43.

[14] MALETZKY P, BAUER R, LAHNSTEINER J, et al. 1999 Immobilisation of iron ions on nafion® and its applicability to the photo-fenton method. *Chemosphere*, 38(10): 2315-25.

[15] FENG J, HU X, YUE P L, et al. 2003 A novel laponite clay-based Fe nanocomposite and its photo-catalytic activity in photo-assisted degradation of Orange II. *Chemical Engineering Science*, 58(3–6): 679-85.

[16] LE N M A D, CASTIGLIONI J, BUSSI J, et al. 2008 Catalytic activity of an iron-pillared montmorillonitic clay mineral in heterogeneous photo-Fenton process. *Catalysis Today*, s 133–135(1): 600-5.

[17] BOZZI A, YURANOVA T, MIELCZARSKI J, et al. 2002 Abatement of oxalates catalyzed by Fe-silica structured surfaces via cyclic carboxylate intermediates in photo-Fenton reactions. *Chemical Communications*, 19(19): 2202-3.

[18] RIOS-ENRIQUEZ M, SHAHIN N, DUR N-DE-BAZ A C, et al. 2004 Optimization of the heterogeneous Fenton-oxidation of the model pollutant 2,4-xylidine using the optimal experimental design methodology. *Solar Energy*, 77(5): 491-501.

[19] LV X, XU Y, LV K, et al. 2005 Photo-assisted degradation of anionic and cationic dyes over iron(III)-loaded resin in the presence of hydrogen peroxide. *Journal of Photochemistry & Photobiology A Chemistry*, 173(2): 121-7.

[20] TRYBA B. 2008 Immobilization of TiO2 and Fe-C-TiO2 photocatalysts on the cotton material for application in a flow photocatalytic reactor for decomposition of phenol in water. J Hazard Mater. *Journal of hazardous materials,* 151(2-3): 623-7.

[21] REIJNDERS L. 2006 Cleaner nanotechnology and hazard reduction of manufactured nanoparticles . *Journal of Cleaner Production*, 14(2): 124-33.

[22] AUFFAN M, ACHOUAK W, ROSE J, et al. 2008 Relation between the redox state of iron-based nanoparticles and their cytotoxicity toward Escherichia coli. *Environmental Science & Technology*, 42(17): 6730-5.

[23] KELLER A A, GARNER K, MILLER R J, et al. 2012 Toxicity of Nano-Zero Valent Iron to Freshwater and Marine Organisms . *Plos One*, 7(8): 1115-9.

[24] LI Z Q, GREDEN K, ALVAREZ P J J, et al. 2010 Adsorbed polymer and NOM limits adhesion and toxicity of nano scale zerovalent iron to E. coli. *Environmental Science & Technology*, 44(9): 3462-7.

From Centralized Disassembly to Life Cycle Management: Status and Progress of E-waste Treatment System in China

Xiaolong Song[1,2], Jianxin Yang[3], Bin Lu[3] and Dong Yang[4]

[1]WEEE Research Centre of Shanghai Polytechnic University, 201209 Shanghai, China
[2]Shanghai Collaborative Innovation Centre for WEEE Recycling, 201209 Shanghai, China
[3]State Key Laboratory of Urban and Regional Ecology, Research Center for Eco-Environmental Sciences, Chinese Academy of Sciences, 100085 Beijing, China
[4]SDAS key laboratory of scientific decision-making, Institute of science and technology for development of Shandong province, Shandong Academy of Sciences, 250014 Jinan, China

E-mail: songxiaolong@sspu.edu.cn

Abstract. China is now facing e-waste problems from both growing domestic generation and illegal imports. Many stakeholders are involved in the e-waste treatment system due to the complexity of e-waste life cycle. Beginning with the state of the e-waste treatment industry in China, this paper summarizes the latest progress in e-waste management from such aspects as the new edition of the China RoHS Directive, new Treatment List, new funding subsidy standard, and eco-design pilots. Thus, a conceptual model for life cycle management of e-waste is generalized. The operating procedure is to first identify the life cycle stages of the e-waste and extract the important life cycle information. Then, life cycle tools can be used to conduct a systematic analysis to help decide how to maximize the benefits from a series of life cycle engineering processes. Meanwhile, life cycle thinking is applied to improve the legislation relating to e-waste so as to continuously improve the sustainability of the e-waste treatment system. By providing an integrative framework, the life cycle management of e-waste should help to realize sustainable management of e-waste in developing countries.

1. Introduction

E-waste (also called WEEE) is one of the fastest growing waste streams [1]. E-waste not only contains abundant amounts of recyclable materials but also hazardous substances, e.g. heavy metals, polybrominated biphenyls (PBBs), and polybrominated diphenyl ethers (PBDEs). With the increasing consumption and turnover of e-products, China is now facing the dual pressures of domestic generation and illegal transboundary movement of e-waste [2,3].

E-waste treatment system in China has been through several phases from informal family workshops to licensed plants with specialized subsidy. Prior to 2003, e-waste treatment system was spontaneous family workshops. Early 2004, the National Development and Reform Commission (NDRC) determined Zhejiang province and Qingdao city as the pilot province and city for the construction of a national e-waste treatment system [4]. In June 2009, the Chinese government released the "*Household Appliances Old for New Rebate Program*" (Old for New Program) in five large cities and four provinces and then extended to most of other provinces. The "Old for New

Program" ended on 31 December 2011. From July 2012, WEEE Treatment Fund Policy (Fund Policy) was implemented, became the new diving force for promoting Chinese e-waste treatment system. Since implementation of the "Old for New Program" and especially the "Fund Policy", the formal treatment system of e-waste has been improved dramatically.

Life cycle management (LCM) refers to the comprehensive management of the whole of the life cycle of products and service for the purpose of realizing sustainable production and consumption [5]. Apart from enterprise and business development, its concepts and methods can also be used to guide the management of solid waste [6–8]. As the largest country exporting e-products and illegally importing e-waste, China plays an important role in the life cycle management of the global electronic industry [9, 10].

Beginning with the state of the e-waste treatment industry in China, this paper first analyzes the industry's scale, treatment capacity, formal plant types, informal sector transition, etc. Then, the latest development in e-waste management are analyzed (including the new edition of the China RoHS Directive, new Treatment List, new funding subsidy standard, and eco-design pilots for e-products). Finally, from the perspective of the whole process, this paper proposes a conceptual model for the life cycle management of e-waste. Moreover, the research elaborates the specific strategies and prospects for performing LCM, and attempts to provide a new methodology for optimizing e-waste treatment in China so as to realize sustainable management.

2. Development of e-waste treatment industry

2.1. Treatment capacity of e-waste

Large quantities of e-products in China have been discarded in recent years. With progress in technology and growth in living standards, the trend in e-products discard is an indication of the rapid increase in consumption and speed of update. As a consequence, the quantity of e-waste discarded increases year-on-year. As an example, the number of televisions, refrigerators, washing machines, air conditioners, and personal computers discarded has grown from 51.48 million in 2009 to 113.78 million in 2014 [11], corresponding to an average annual growth rate of 24.2%.

With the implementation of its "Fund Policy", the Ministry of Environmental Protection (MEP) has begun to count up the numbers of televisions, refrigerators, washing machines, air conditioners, and personal computers treated by licensed plants since July 2012. According to these statistics, 7.68, 39.87, and 70.2 million of these five kinds of e-waste have been formally treated in consecutive years from 2012 to 2014. Correspondingly, the formal treatment rate has risen from 10.13% to 61.7% (Figure 1).

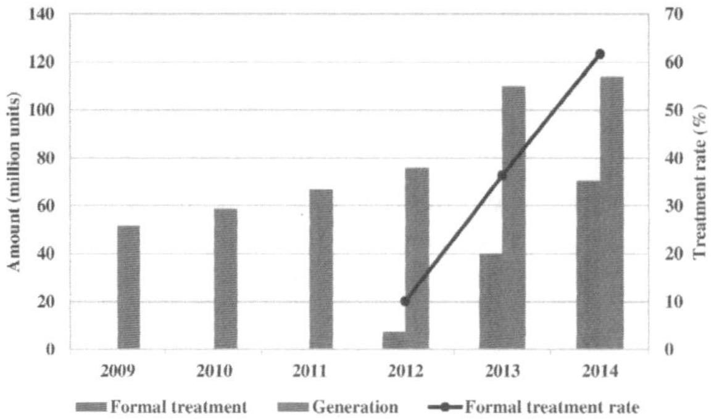

Figure 1. Generation and treatment of typical e-waste in China (2009–2014).

Among the five kinds of e-waste, it is the discarded televisions that are mainly disassembled by the licensed plants at present. For instance, in 2013, 92.0% of the total amount of the e-waste formally disassembled by licensed plants comprised of discarded televisions, while the other four kinds made up 8.0% in total [12]. In 2014, the proportion attributable to discarded televisions decreased slightly. Televisions, refrigerators, air conditioners, washing machines, and personal computers disassembled occupied 81.8%, 2.24%, 0.18%, 4.68%, and 11.1% of the total disassembled amount, respectively. The reasons why discarded televisions are the main kind of object treated are as follows. On the one hand, licensed plants are enthusiastic about disassembling televisions due to the high subsidy involved. On the other, the other kinds of e-waste, e.g. air conditioners and washing machines, have relatively high collection costs and so it is uneconomical for licensed plants to purchase them from collectors. As a result, most of them are processed in informal sector.

2.2. *Structure of the formal sector*

The formal e-waste treatment plants in China can be split into four types. The first one consists of e-waste recyclers that have received WEEE treatment license. E-waste recyclers registered on the e-waste disassembly plant list form the second one. The third consists of recyclers authorized for hazardous e-waste recycling and disposal. The fourth kind is those engaged in the treatment of imported category-seven solid waste. The four categories are not clear-cut: there is crossover between these four kinds of licenses and some formal plants generally present multiple qualifications at the same time.

Since the "Fund Policy" came into effect, the first kind of formal treatment plant has become the main force involved in the formal treatment of e-waste. By the end of 2015, the Ministry of Finance (MF) had announced that 109 formal e-waste treatment plants qualified for funding subsidy. These plants are located in 29 provinces, but are mainly distributed in central and eastern China (Figure 2). With the rapid increase and comprehensive coverage afforded by these licensed treatment plants, the layout of the formal sector for treating e-waste in China has been preliminarily completed.

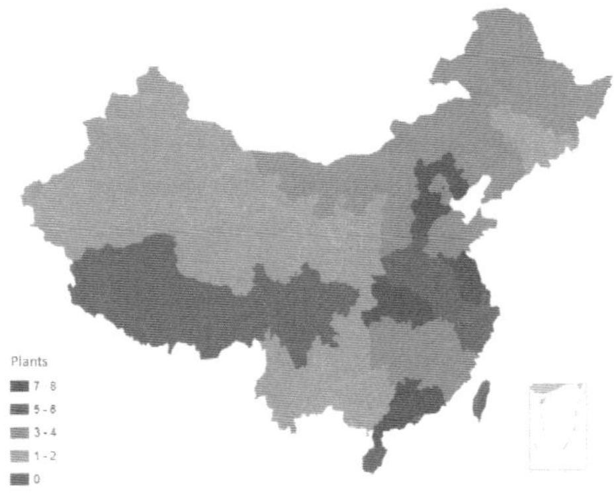

Figure 2. Distribution of formal WEEE treatment plants in China.

2.3. *Transition of the informal sector*

A key issue for China's e-waste management is how to set up incentives for informal sector so as to reduce improper recycling activities and to divert more e-waste flow into the formal sector [13]. Guiyu, a small town located in southeastern China, is known across the world for polluting the environment as a result of its illegal e-waste disassembly [14]. As a consequence, this region has become a typical representative of the informal sector for recycling e-waste. Since the middle of 1990s, farmers living in Guiyu, initiated e-waste recycling activities [15]. However, these informal workshops adopted primitive methods such as open burning, open dumping, and acid washing, caused serious environment pollution and threatened human health [16].

In October 2005, Guiyu was included in the first group of circular economy pilot projects for recycling waste household appliances. In December 2010, Guiyu established a circular economy park with the theme of e-waste disassembly and treatment, hoping to transform informal recycling activities to formal ones through centralized disassembly and pollution treatment. By 2015, construction of the park had made substantial progress: of 5,169 disassembly households, 2,028 had been banned. Furthermore, 29 companies (composed of 758 modified disassembly households from the 3,141 rectified households) had entered the zone to perform formal disassembly and treatment [17].

3. Latest progress in e-waste management

In order to deal with the environmental problems of e-waste and improve the e-waste treatment industry, the Chinese government has been issued a variety of environmental laws, regulations, standards, and norms related to e-products production and e-waste management [18].

As the counterpart of the EU's RoHS and WEEE Directive, China passed the "Ordinance on Management of Prevention and Control of Pollution from Electronic and Information Products" in 2006 and the "Administration Regulation for the Collection and Treatment of Waste Electronic and Electrical Equipment" in 2009. Generally, the two documents are unofficially referred to as China RoHS Directive and China WEEE Directive [19].

3.1. China RoHS Directive (2016 edition)

On 6 January 2016, eight ministries and commissions, including the Ministry of Industry and Information Technology (MIIT), published "The Restriction of Hazardous Substances in Electrical and Electronic Equipment". It supersedes the document "Ordinance on Management of Prevention and Control of Pollution from Electronic and Information Products" to be the new and current edition of China RoHS Directive.

The new edition of RoHS Directive proposes the compilation of a management catalogue to restrict the use of hazardous substances. To be more specific, it stipulates the categories of e-products of concern, the types of hazardous substances restricted, and any exceptional requirements.

In addition, the new edition of RoHS Directive adds a system of conformity assessment to the restriction of the use of hazardous substances contained in e-products. For the e-products contained in the compliance management catalogue, the amount of hazardous substances used is supposed to satisfy the national or industry standard and the products have to be managed according to the conformity assessment system.

3.2. WEEE Treatment List (2014 edition)

To help coordinate the implementation of China WEEE Directive, a WEEE Treatment List (first batch) was released in 2011, which included five kinds of e-waste: televisions, refrigerators, washing machines, air conditioners, and personal computers. On this basis, the licensed plants were to mainly disassemble these five kinds of e-waste.

As the e-waste treatment industry in China has developed, an expansion of the scope of the treatment catalogue has become a realistic demand. In 2013, the NDRC and MF jointly evaluated the implementation of the first batch of the treatment catalogue and studied if adjustments to the catalogue were necessary. After extensive consultation and research, the WEEE Treatment List (2014 edition) was officially released in February 2015. In the new edition, e-waste has been extended to 14 categories based on the original 5 types. The added categories are range hood, electric water-heater, gas water-heater, printer, copier, fax machine, monitor, mobile phone and single-machine telephone.

3.3. WEEE fund standard (2016 edition)

According to "Fund Policy", the e-products that are sold and used in China are the objectives for the fund collection. So the domestic producers and consignees or agents of imported e-products should fulfill their payment obligations of the fund. Moreover, the imported units are included for the fund collection while the exported ones are excluded [20]. Currently, the producer and importer of e-products should pay 13 yuan Renminbi (RMB), 12 RMB, 7 RMB, 7 RMB and 10 RMB, for each television, refrigerator, washing machine, air conditioner and personal computer, respectively.

By keeping an audit of e-waste disassembly, the licensed plants can acquire the fund. The MF started to adjust the existing funding subsidy in 2015 according to changes in the treatment costs and the profits from e-waste. The newly established subsidy standard was put into force on 1 January 2016. Compared with the old standard (formulated in 2012), the new 2016 standard are predominantly characterized by a reduction in the subsidy available for disassembling televisions and personal computers and an increase in the subsidy for disassembling air conditioners (Table 1).

Table 1. WEEE funding subsidy standard in China.

Category	2012 standard (RMB yuan)	2016 standard (RMB yuan)
Television	85	60-70
Refrigerator	80	80
Air conditioner	35	130
Washing machine	35	35-45
Personal computer	85	70

3.4. Eco-design of e-products

Apart from strengthening the supervision of e-waste disassembly and recycling processes, the Chinese government has begun to control the pollution sources and to utilize resources more effectively from the perspective of eco-design. Since July 2014, the MIIT has begun to establish eight demonstration industries to coordinate eco-design of industrial products. The aim is to transform industrial pollution control from end-of-pipe treatment to whole life cycle management.

The first batch of eco-design pilot enterprises was published in June 2015. It included seven e-products producers mainly producing televisions, liquid crystal displays (LCDs), refrigerators, air conditioners, washing machines, personal computers, dust collectors, air cleaners, relays, lead-acid batteries, etc. At present, the second batch of eco-design pilot enterprises is under declaration and more enterprises are supposed to participate in this work.

3.5. Extended producer responsibility in e-products areas

In June 2015, the MIIT launched a pilot scheme on extended producer responsibility (EPR) in electronic industry. The aim was that the producers should play a leading role in the design, production, collection and recycling of e-products.

The pilot work involved household appliances, computers and office electrical appliances, communication and electronic products, lighting electrical appliances, and batteries. In addition, a total of 15 enterprises involved in manufacturing electronic products were considered in this pilot project.

4. Future perspective: life cycle management of e-waste

To improve e-waste recycling efficiency and effectively control environmental risk, a conceptual model for the life cycle management of e-waste is proposed (Figure 3).

When considering the life cycle stages, it is necessary to obtain relevant information about key components, material metabolism, and input/output of materials in the treatment process, starting from its generation to the final disposal of the e-waste. The life cycle implications of e-waste can then be systematically analyzed and diagnosed using appropriate tools including material flow analysis (MFA), LCA, life cycle costing (LCC), and social life cycle assessment (S-LCA). This leads to the decision-making process. In the process, it is helpful to consider life cycle engineering aspects including eco-design, recycling techniques, and prevention and control of pollution. Finally, the life cycle thinking stage is reached wherein consideration is given to laws, regulations, standards, and government policies in place.

Based on life cycle thinking and EPR principle, the "Fund Policy" implemented in China requires the producers and importers of e-products to meet the cost of treat e-waste. At present, the "Fund Policy" is still in its infancy, a lot of endeavor is needed to improve its optimization [20]. Depending on the effect of the "Fund Policy", a dynamic regulation mechanism needs to be established in the future to adjust the fund collected and subsidy standard so as to improve the policy performance. The fund collected for treating e-waste inevitably increases the economic burden on producers. For e-products enterprises adopting eco-design or participating in e-waste recycling, the fund collected is meant to be reduced within a certain range from a life cycle perspective. On the pretext of maintaining the stability of the total fund, the funding subsidy standard is supposed to be adjusted according to development in the industry in a timely manner.

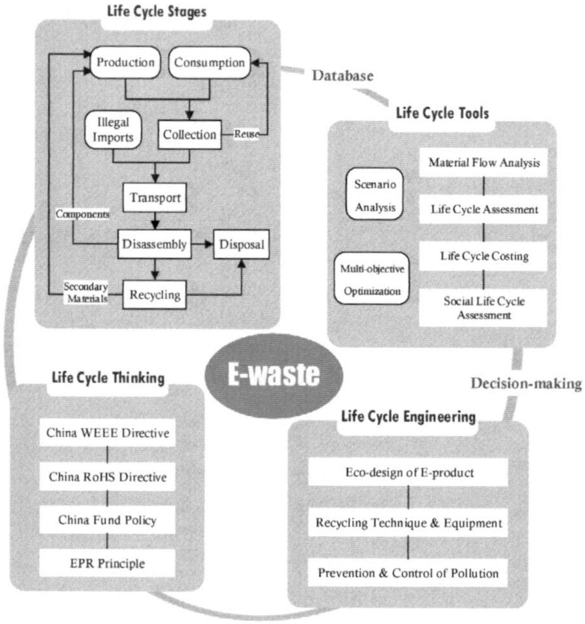

Figure 3. Life cycle management scheme for e-waste.

5. Conclusions

There are large amounts of e-waste generated in China which have a high recycling value. In recent years, the amount of e-waste formally treated has increased at a steady rate. As the number of formal treatment plants for e-waste has increased, the layout of the formal sector has been basically realized providing an increasingly enhanced recycling ability. Meanwhile, by enhancing environmental supervision and effectively controlling illegal disassembly activities, the informal sector have been transformed to formal one.

To implement an LCM scheme for e-waste, four aspects need to be considered simultaneously: the life cycle stages, life cycle tools, life cycle engineering, and life cycle thinking. Our suggestion is to first identify the life cycle stages of the e-waste and extract the important life cycle information. Then, life cycle approaches and tools can be used to conduct a systematic analysis to help decide how to maximize the benefits from a series of life cycle engineering processes. Meanwhile, life cycle thinking is applied to improve the laws and regulations, standards, and policies relating to e-waste so as to continuously improve the sustainability of the e-waste treatment system. This framework should help to realize sustainable management of e-waste in developing countries where legislation surrounding e-waste management is imperfect and the treatment system is relatively immature.

Acknowledgements

The authors gratefully acknowledge the financial supports from the National Natural Science Foundation of China (No.41501598), Innovation Program of Shanghai Municipal Education

Commission (No.15ZZ102) and Science & Technology Planning Project of Shandong Province, China (No.2014GGC01070). The authors also thank the Shanghai Collaborative Innovation Centre for WEEE Recycling for funding and support.

References

[1] Manomaivibool P and Vassanadumrongdee S 2011 Extended Producer Responsibility in Thailand: Prospects for Policies on Waste Electrical and Electronic Equipment *Journal of Industrial Ecology* 15(2) 185-205
[2] Yang J X, Lu B and Xu C 2008 WEEE Flow and Mitigating Measures in China *Waste Management* 28(9) 1589-1597
[3] Zhou L and Xu Z M 2012 Response to Waste Electrical and Electronic Equipments in China: Legislation, Recycling system, and Advanced Integrated Process. *Environmental Science & Technology* 46(9) 4713-4724
[4] Li J H, Yang J and Liu L L 2015 Development potential of e-waste recycling industry in China *Waste Management & Research* 33(6) 533-542
[5] UNEP 2006 Guide to Life Cycle Management—A Bridge to Sustainable Products
[6] Fullana P P, Puig R, Bala A, et al. 2011 From Life Cycle Assessment to Life Cycle Management: A Case Study on Industrial Waste Management Policy Making *Journal of Industrial Ecology* 15(3) 458-475
[7] Song X L, Yang J X, Lu B, et al. 2013 Exploring the life cycle management of industrial solid waste in the case of copper slag *Waste Management & Research* 31(6) 625-633
[8] Weitz K, Barlaz M, Ranjithan R, et al. 1999 Life Cycle Management of Municipal Solid Waste. *The International Journal of Life Cycle Assessment* 4(4) 195-201
[9] Yu J L, Williams E, Ju M T, et al. 2010 Managing E-waste in China: Policies, Pilot Projects and Alternative Approaches *Resources, Conservation and Recycling* 54(11) 991-999
[10] Wang F, Kuehr R, Ahlquist D, et al. 2013 E-waste in China: A Country Report *United Nations University*
[11] CHEARI 2015 White Paper on Chinese WEEE Recycling Industry in China (2014)
[12] Tang A J and Wan L 2014 Discussion on the development and trend of dismantling industry of waste electrical and electronic equipment *Renewable Resources and Circular Economy* 7(3) 21-24 (in Chinese)
[13] Chi X W, Streicher-Porte M, Wang M, et al. 2011 Informal Electronic Waste Recycling: A Sector Review with Special Focus on China *Waste Management* 31(4) 731-742
[14] BAN 2002 Exporting harm: the high-tech trashing of Asia *Basel Action Network*
[15] He Y X and Xu Z M 2014 The status and development of treatment techniques of typical waste electrical and electronic equipment in China: A review *Waste Management & Research* 32(4) 254-269
[16] Herat S and Agamuthu P 2012 E-waste: a problem or an opportunity? Review of issues, challenges and solutions in Asian countries *Waste Management & Research* 30 1113-1129
[17] EID 2015 The development of Guiyu circular economy park Availabled from: http://www.ezaisheng.com/news/show-32991.html
[18] Zeng X L, Li J H, Stevels A L N, et al. 2013 Perspective of Electronic Waste Management in China based on a Legislation Comparison between China and the EU *Journal of Cleaner Production* 51 80-87
[19] Ongondo F O, Williams I D and Cherrett T J 2011 How are WEEE doing? A global review of the management of electrical and electronic wastes *Waste Management* 31(4) 714-730
[20] Yu L L, He W Z, Li G M, et al. 2014 The development of WEEE management and effects of the fund policy for subsidizing WEEE treating in China *Waste Management* 2014 34(9) 1705-1714

Simulation and Experimental Study on Cavitating Water Jet Nozzle

Wei Zhou [1,3], Kai He* [1,2], Jiannan Cai [1,3], Shaojie Hu [1,4], Jiuhua Li [3] and Ruxu Du [5]

[1] Shenzhen Institutes of Advanced Technology, Chinese Academy of Sciences, Shenzhen, China
[2] Shenzhen Key Laboratory of Precision Engineering, Shenzhen, China
[3] China University of Petroleum, Beijing, China
[4] University of South China, Hengyang, China
[5] The Chinese University of Hong Kong, Hong Kong, China

*Email: kai.he@siat.ac.cn

Abstract. Cavitating water jet technology is a new kind of water jet technology with many advantages, such as energy-saving, efficient, environmentally-friendly and so on. Based on the numerical simulation and experimental verification in this paper, the research on cavitating nozzle has been carried out, which includes comparison of the cleaning ability of the cavitating jet and the ordinary jet, and comparison of cavitation effects of different structures of cavitating nozzles.

1. Introduction

The cavitating jet is a kind of water jet technology promoting the rapid evaporation of water and gas evolution in the water jet flow artificially by changing structure of the nozzle, reducing local pressure below the saturating vapor pressure of the liquid [1]. The effect of the jet is enhanced by the strong impact force generated by the bubble burst, shown in Figure 1. At present, the erosion mechanism of the cavitation jet is generally regarded as the theory of the mechanical action. According to the mechanical action theory, cavitation erosion is caused by the strong shock of the micro jet and microwave generated by bubble collapse [2].

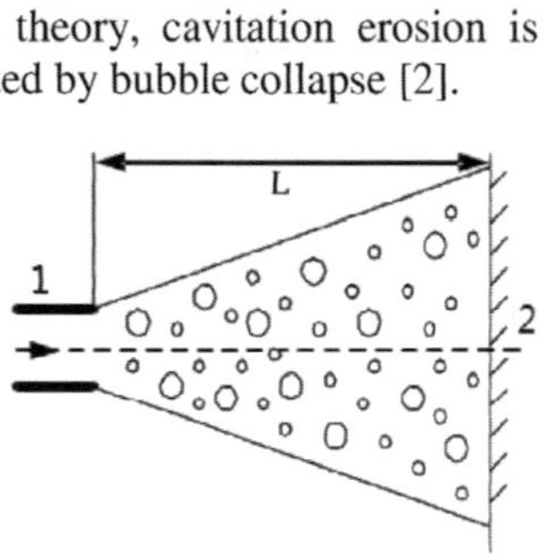

Figure 1. Schematic diagram of cavitating jet (1 is nozzle. L is target distance. 2 is target surface).

Hammit [3]concluded by calculation and experimental measurement that when the cavitation bubble collapses, micro jet velocity near the wall reaches 70~180m/s, the pressure at the surface can up to 140MPa, the frequency of impact at surface from micro jet is about 100~1000 / (s·cm²).

Therefore, the cavitating jet has superior cleaning ability without very high pressure, and also can avoid damage to the surface coving expensive paint or anticorrosive coating. In addition, it can be used underwater, not like high pressure water jet weakening dramatically by environmental media. To the same cleaning efficiency, the pressure of cavitating jet is only 1/4 of the high pressure water jet, and the maximum does not exceed 16MPa [4].

The current cavitating jet is mainly divided into two types： artificial submerged jet created by Vijav [5] and the self-resonance cavitating jet proposed by Johnson V E [6] and Conn A F [7]. The former improves on the basis of submerged jet, under which cavitating water jet can work better [5]. The latter generates cavitating jet by resonant cavity. This paper focuses on the self-resonace cavitating jet under the submerged condition. The organ pipe nozzle (shown in Figure 2) and Helmholtz nozzle are the most common two kinds of structures. With the further research on cavitation， Liao Z F [8] invented a new nozzle with better cavitation effect on the basis of the Helmholtz nozzle, a self-resonance pulsed nozzle (shown in Figure 3). This paper gives the comparison research on the two kinds of nozzles through experiment and simulation based on the application of underwater fouling organism cleaning.

2. Structure of nozzle and mesh generation

As shown in Figure 2, the main structural parameters of Nozzle 1 include diameter D_s and length h of inlet, diameter D and length L of resonant cavity, diameter D_w and length H of outlet. The main structural parameters of Nozzle 2 include diameter d_1 of upstream nozzle, diameter d_2 of downstream nozzle, diameter D_c of self-resonance cavity, length L_c, collision angle α as shown in Figure 3.

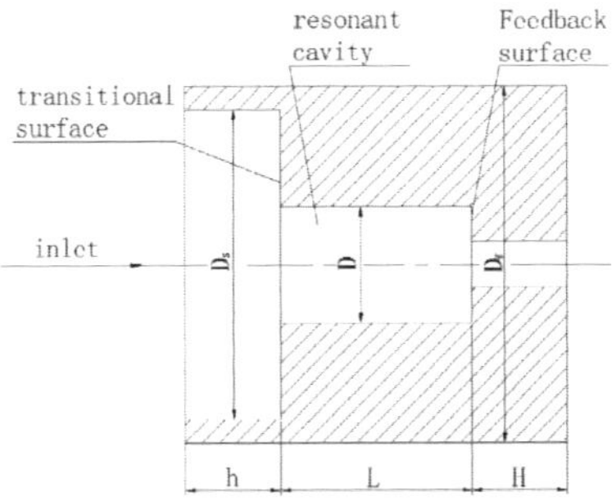

Figure 2. Nozzle 1: organ pipe nozzle.

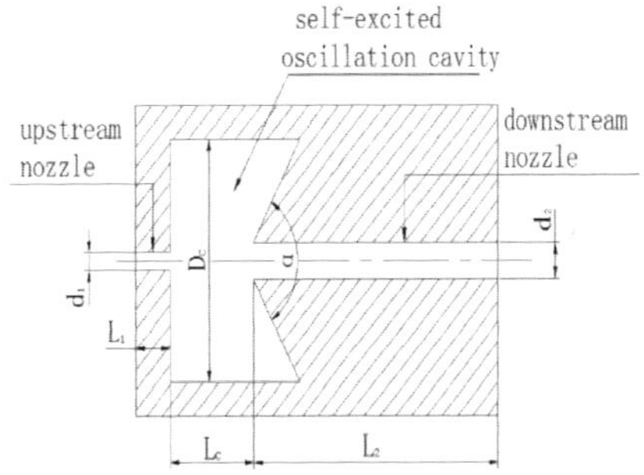

Figure 3. Nozzle 2: self-resonance pulsed nozzle.

By using SolidWorks software, CAD models of Nozzle 1&2 are established, then are meshed by ICEM. Grid model of Nozzle 1&2 as shown in Figure 4 and Figure 5. Mesh quality is above 0.7. Grid numbers are 889602 and 213948， absolutely ensuring the accuracy of the simulation.

Figure 4. Grid model of Nozzle 1.

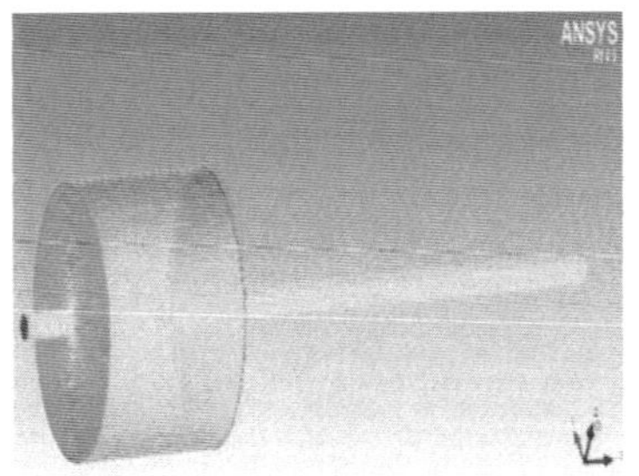

Figure 5. Grid model of Nozzle 2.

3. Numerical simulation and model selection

As water is incompressible viscous fluid, its motion law is controlled by Navier-Stokes equations, as in Equation (1) and Equation (2)：

$$\frac{\partial u_i}{\partial x_i} = 0 \tag{1}$$

$$\frac{\partial u_i}{\partial t} + u_i \frac{\partial u_i}{\partial x_j} = -\frac{1}{\rho}\frac{\partial P}{\partial x_i} + \frac{\partial}{\partial x_j}\left(\vartheta \frac{\partial u_j}{\partial x_j}\right) \tag{2}$$

Where u is the velocity of fluid, P is the pressure of fluid, ρ is the density of fluid and ϑ is the kinematic viscosity.

By using FLUENT software, choosing the mixture model as multiphase model, flow model is established on the basis of Navier-Stokes equations. Turbulent model uses RNG k-ε model [9]. Compared with standard k-ε model, it is usually considered as turbulence vortex， so it has higher accuracy and reliability. Cavitation model uses Schnerr-Sauer model [10]. Parameter of flowing medium uses fluent database parameters. In the computation, the inlet of the mixing flow uses the inlet pressure boundary condition with 15MPa, and the outlet of the mixing flow uses the pressure boundary condition with one bar.

4. Simulation Results & Analysis

4.1. Simulation Results of nozzle 1

As shown in Figure 6, the cavitating region of Nozzle 1 is mainly concentrated in small areas on both sides of the outlet. The vorticity figure also shows that there are more intense vortexes, so it is easy for cavitation to occur. From the pressure diagrams, cavitation occurs because of the pressure which is lower than the saturated water vapor pressure. The highest gas fraction of the cavitation region is 0.216.

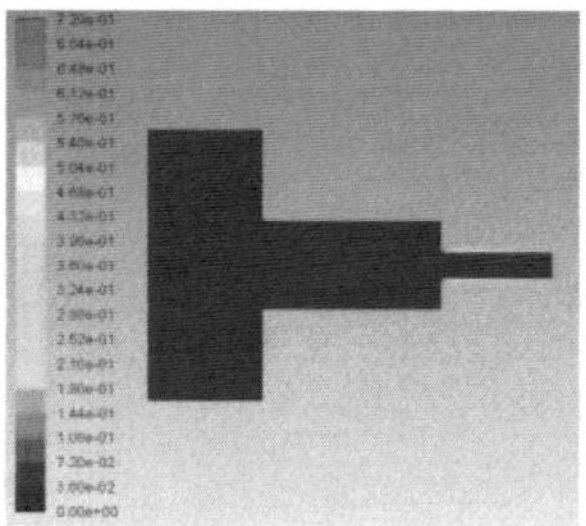

(a) contours of volume fraction (vapor)

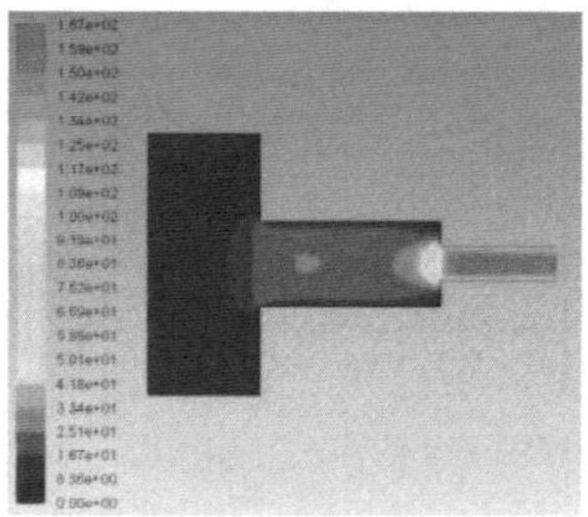

（b）contours of velocity (m/s)

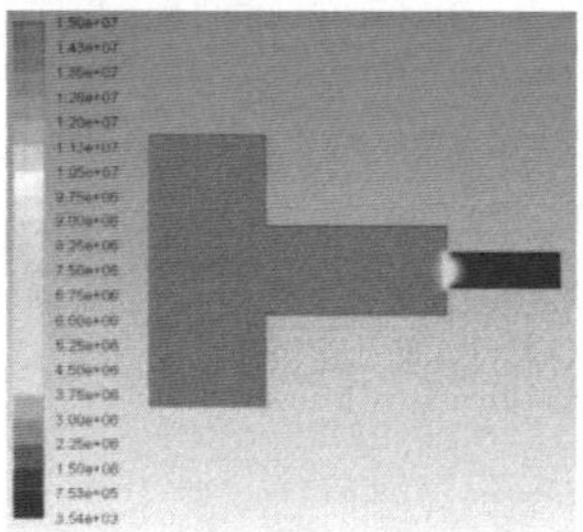

(c) contours of pressure (mixture) (pascal)

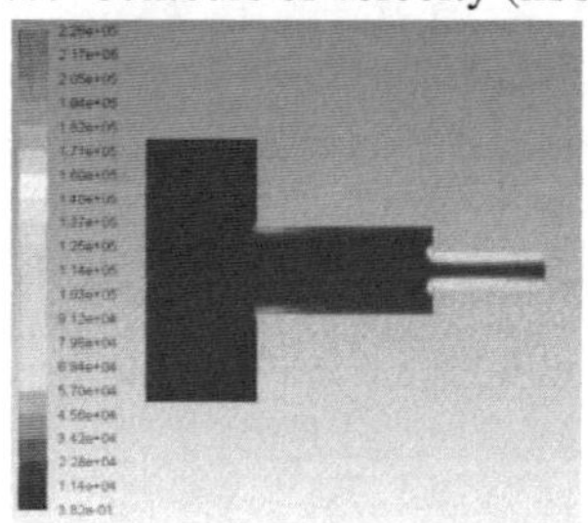

（d）contours of vorticity(1/s)

Figure 6. Contours of nozzle 1.

4.2. Simulation Results of nozzle 2

It can be seen from the simulation results in Figure 7 that the cavitation area of Nozzle 2 is larger and low pressure area is also larger and cavitation phenomenon is stronger. In the middle of a resonant cavity, the most volume fraction of vapor ups to 0.9.Certainly, the average gas fraction at the outlet is also above 0.286, larger than Nozzle 1. So, Nozzle 2 is more suitable for cavitation.

(a) contours of volume fraction (vapor)

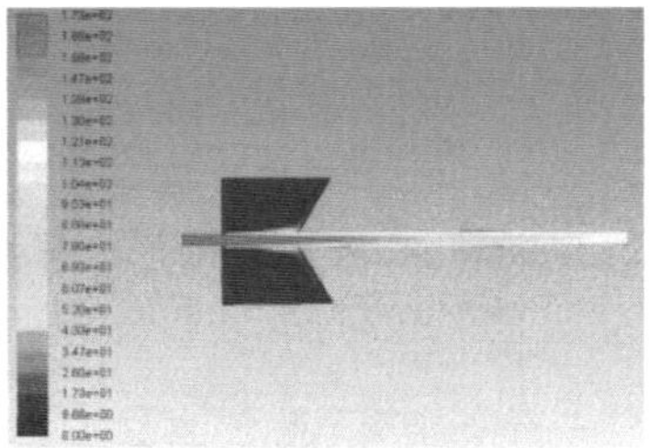

（b）contours of velocity (m/s)

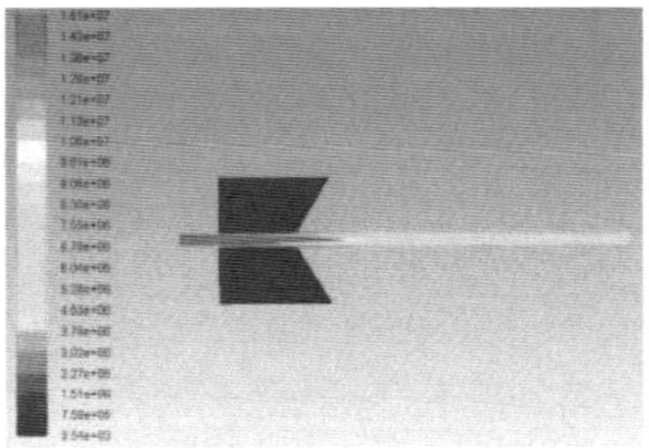

(c) contours of pressure (mixture) (pascal)

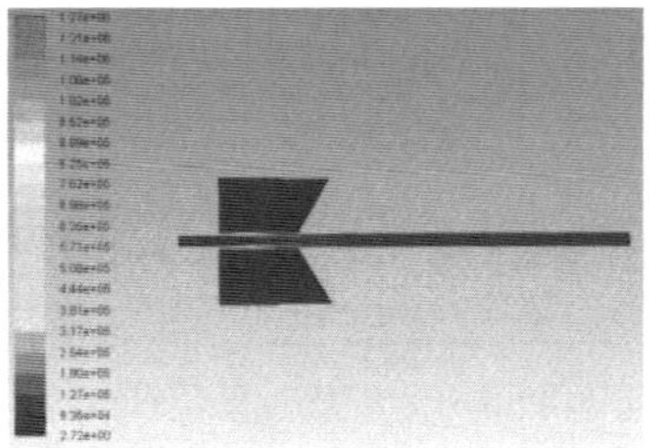

（d）contours of vorticity(1/s)

Figure 7. Contours of nozzle 2.

5. Experiments

5.1. Experimental device

Experimental device is shown in Figure 8 and Figure 9. The water supply system is composed of a water tank and a water pump. In order to reach the submerged condition, the cleaning experiment was carried out in the sealed chamber. The nozzles clean the sample vertically, keeping target distance 20mm. The structure of various nozzles are shown from Figure 10 to Figure 12 respectively.

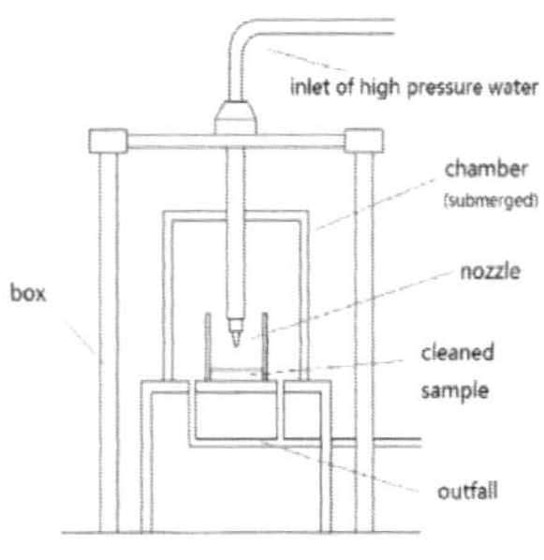

Figure 8. Schematic diagram of experimental device.

Figure 9. Experimental device.

Figure 10. Common nozzle.

Figure 11. Nozzle 1.

Figure 12. Nozzle 2.

5.2. Derusting experiment

In order to compare the difference of striking forces caused by cavitating jet and ordinary jet, we use the nozzle 1 and conventional nozzle to do rust experiment. The pressure of inlet water is 15MPa. The cleaning effect is shown in Figure 13. The experiment result indicates that under a water pressure of 15MPa, cavitating jet can remove rust but conventional jet cannot. We can conclude that the striking force of cavitation jet is larger than conventional jet under the same pressure.

(a) pre-derusting (b) post-derusting

Figure 13. The contrast diagram of derusting.

5.3. Fouling organism cleaning experiment

Fouling organism mainly includes barnacles, shellfish and so on. We used the stones covered with mussels, barnacle as experimental sample to verify the feasibility of cavitating jet in the cleaning application and to compare the effect of the two kinds of cavitating nozzles. In this comparative experiments between the nozzle 1 and nozzle 2, the inlet pressure is 12.5MPa, and the target distance is 20mm and the cleaning time is 10s. Shown in Figure 14 and Figure 15, the cleaning area of Nozzle 1 is 10cm × 2cm and that of Nozzle 2 is 14cm × 2.5cm.

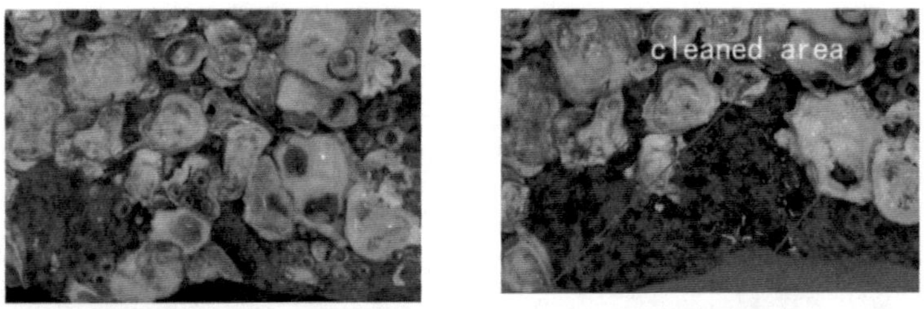

(a) pre-cleaning (b) post-cleaning

Figure 14. The cleaning effect of Nozzle 1.

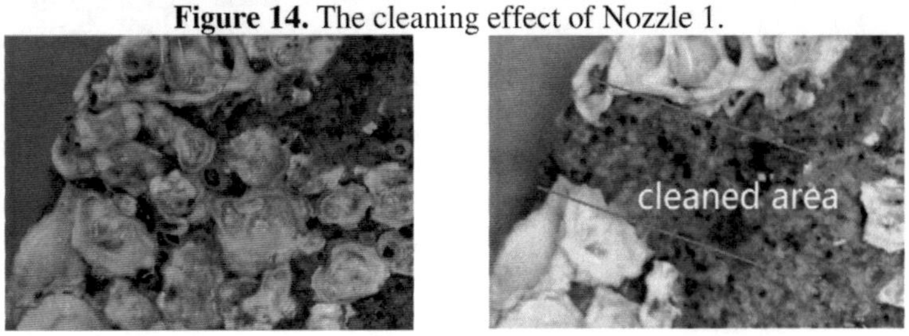

(c) pre-cleaning (d) post-cleaning

Figure 15. The cleaning effect of Nozzle 2.

From the figure, it can be seen clearly that both the nozzle 1&2 have a good cleaning effect. But, the cleaning area of jet of Nozzle 2 is wider than that of Nozzle 1. So the cleaning efficiency of the Nozzle 2 is higher than that of the Nozzle 1.

6. Conclusions

This paper has studied the cavitating water jet nozzle by simulation and experiment. We believe that cavitation water jet, especially generated by self-resonance pulsed nozzle will has broad application prospects in the field of cleaning. The conclusions obtained are as follows.

- According to the simulation results, in the organ pipe nozzle, cavitation occurred mainly in the mutational cross-section of the position, especially in the outlet which is the narrow area with the lower pressure, so vapor is more. In the self-resonance pulsed nozzle, cavitation occurred not only in the same regions above, but also in resonant cavity, due to more vortex flow where absolute pressure is lower.
- Based on the simulation results, the cavitation in the self-resonance nozzle is more serious than that in the organ pipe nozzle. The gas fraction is higher and the cavitation zone is larger.
- The experimental results verify the hitting force of cavitating jet is greater than the ordinary in same pressure, and the former is even effective to remove rust in 15MPa. In addition, the cleaning effect of the self-resonance nozzle is better than that of the organ pipe nozzle.

Acknowledgments

This work was accomplished under the financial support provided by Shenzhen Technology Development project (CXZZ20140904104336050).

References

[1] Peng Guoyi, Shimizu Seiji 2013 Progress in numerical simulation of cavitating water jets. *Journal of Hydrodynamics*, 04:502-509.

[2] Yang Chunming 2014 *The Experimental Research of Metal Surface Strengthening of Jet Cavitation Peening* . China University of Mining and Technology.

[3] Hammit F G 1980 *Cavitation and multiphases flow phenomena.* New York: Mc Graw-Hill Book Co.

[4] Zhang Lixia 2006 Sea biological fouling prevention technology. *Rubber & Plastics Resources Utilization*, 02:7-11+14.

[5] Vijay M M, et al 1992 A study of the practicality of cavitating water jets. proc, 11th Int. Symp. *On jet cutting technology*. St. Andrews, Scotland.

[6] Johnson V E, et al 1982 Self-resonating cavitating jets. Proc.6th Int, Symp. *On water jet technology*, England.

[7] Conn A F, Johnson V E. 1982 *Fluid dynamics of submerged cavitating jet cutting jet for deep hole bits.* SPE11060.

[8] Liao Z F, Tang C L 2003 Theory and experimental study of the self-excited oscillation pulsed jet nozzle. *Chinese Journal of Mechanical Engineering*, 04:379-383.

[9] Yakhot V, Orzag S A 1986 Renormalization group analysis of turbulence:basic theory. *Journal of Scientific Computing*. 1:3-11.

[10] Schnerr G H, Sauer J. 2001 *Physical and Numerical Modeling of Unsteady Cavitation Dynamics*. In Fourth International Conference on Multiphase Flow, New Orlenas, USA.

The Inter-provincial Transfer of Transportation Hidden Carbon in China's Provinces

Ping Jiao, Jian Wei Wang and Shuai Zhang[1]

Department of Economics and Management, Chang'an University, Xi'an, Shaanxi, 710064, China

[1]E-mail: cd_zhangshuai@163.com

Abstract. Recognize the characteristics of the inter-provincial transportation hidden carbon transfer scientifically, which is significant to evaluate regional carbon emission correctly and reasonably and make the regional carbon reduction policies. With the China's interregional input-output tables in 2007 and 2010 years, this paper measures the transportation hidden carbon and researches the features of inter-provincial transportation hidden carbon transfer. The results show that: (1) The Shanxi, Hubei and et al in central-western regions is the major importer of carbon emission, and the Tianjin, Jiangsu, Fujian and et al in eastern regions is the net exporter; (2)The transfer size and density of the eastern, central and western regions present the decreasing tendency. The more advanced the economy, the higher the transportation hidden carbon emission. In China, the Beijing-Tianjin-Hebei Economic Band is the most cohesive region about size and density. As the results, the responsibility of provincial transportation carbon emission should be properly adjusted, and regional linkage carbon emission policies should be implemented.

1. Introduction

Climate change has become a global environment challenges together. To reduce carbon emission and develop low-carbon economy have become the consensus of the international community. China is a large country which the level of economic development, industrial structure, and so on is obvious differences between different provinces. In the markets, there exists strong inter-provincial commodity trade and flows, reflecting that the local productions are used by other places. This causes the 'carbon leakage' [1]-[3]. Because the transportation is a service of providing spatial movement from one place to another. With the transfer of goods between provinces, the transportation also presents the characteristic of spatial transfer, thus transportation carbon also has this feature. Therefore, in order to measure regional transport carbon emission accurately, realize the responsibility of transport carbon and make the regional carbon emission policies, we need to cognize the condition of inter-provincial transportation carbon emission transfer.

At present, interregional correlation can influence the redistribution of carbon emission region, pollution transfer and so on obviously. And many scholars are interested in this respect. The researches about this most focus on: (1) the measurement, transfer and influences analysis of the interregional hidden carbon under the trade path[4]-[8]. (2) analysis of specific countries and regions carbon[8]-[13]. Some scholars research the hidden carbon transfer from industry perspective [14]-[16], and the papers of single industry focus on industry, electricity and heat production industry and so on. For the transportation carbon emission, the main research directions are the measurement and evaluation about transportation carbon[17]-[19], spatial variance analysis[16],[20], influence factors analysis and so on[21]-[23]. Most of researches in transportation is analyzing based on the direct carbon emission, which is calculated by the data of energy consumption. However, this method ignores the hidden carbon emission, thus it can't evaluate the regional carbon emission accurately. And few scholars research the inter-provincial transportation hidden carbon transfer in China.

Thus, based on the China's interregional input-output tables, this paper analyses the condition of the inter-provincial hidden carbon transfer on transportation. We hope the results can become the references to adjust the responsibility of transport carbon reduction and make the carbon reduction policies.

2. Hidden carbon transfer calculations

Input-output table is the important part in national economic statistics, reflecting the economic transfers between different industries or different regions, so it can present the characterization of interregional transfer accurately. Therefore, based on the interregional input-output table in 2007 and 2010, this paper accounts transportation hidden carbon transfer.

In interregional input-output model, the relationship of supply and demand in different areas can be expressed as follows:

$$X^{R} = A^{RS} X^{R} + F^{RS} + E^{R} \tag{1}$$

In the formula: X^{R} is the output array in interregional input-output model; A^{RS} is the input coefficient matrix; F^{RS} is the final demand matrix; E^{R} is the exporting array.

Because the inter-provincial input-output model is the imported competition type in 2007 and 2010, and the input is not distinguished between imported products and domestic products. In order to account the provincial transportation hidden carbon emissions separately, we need to remove imported parts contained in input. This paper introduces imported excluding coefficient matrix to eliminate the part of import in the matrix. Imported excluding coefficients are evaluated as follows:

Intermediate demand imported excluding coefficient:

$$\hat{M}_{X_j}^{S} = \frac{M_{X_j}^{S}}{\sum_{R=1}^{30}\sum_{i=1}^{6} X_{ij}^{RS}} \tag{2}$$

Final demand imported excluding coefficient:

$$\hat{M}_{F}^{S} = \frac{M_{F}^{S}}{\sum_{R=1}^{30}\sum_{i=1}^{6} F_{i}^{RS}} \tag{3}$$

After exclude, (1) can be rewritten as:

$$X^{R} = \left(I - \hat{M}_{X_j}^{S}\right) A^{RS} X^{R} + \left(I - \hat{M}_{F}^{S}\right) F^{RS} + E^{R} \tag{4}$$

In the formula: $\left(I - \hat{M}_{X_j}^{S}\right) A^{RS} X^{R}$ is intermediate demand excluding imports, $\left(I - \hat{M}_{F}^{S}\right) F^{RS}$ is the final demand supplied by domestic production.

Before calculating inter-provincial transportation hidden carbon transfer, we need to calculate the carbon emission coefficients first. Carbon emission coefficient means the carbon emissions under per unit of output, including direct carbon emission coefficient and complete carbon emission coefficient. The specific formulas are as follows:

$$D^{R} = c^{R} / x^{R} \tag{5}$$

$$L^{RS} = D^{RS}[I - (I - \hat{M}^{S}_{X_1})A^{RS}]^{-1} \tag{6}$$

$$D^{RS} = \begin{bmatrix} D^{R_1} & 0 & \cdots & 0 \\ 0 & D^{R_2} & \cdots & 0 \\ \vdots & \vdots & \ddots & \vdots \\ 0 & 0 & \cdots & D^{R_{30}} \end{bmatrix} \tag{7}$$

Among them, D^{R_i} is direct carbon emission coefficient in R area, which is the proportion of direct carbon emissions c^{R} with the transport sector gross products x^{R}; L^{RS} is the complete carbon emission coefficient between provinces. Interregional transport hidden carbon matrix T^{RS} is as follows:

$$T^{RS} = L^{RS}[I - \hat{M}^{S}_{F})F^{RS}] \tag{8}$$

3. Analysis of measured results

3.1 Analysis of transfer amount

According to the equation (5)-(7), this paper measures the inter-provincial transportation hidden carbon transfer matrix in 2007 and 2010. On this basis, transportation hidden carbon emission value of transfer-in, transfer-out and net transfer between provincial regions can be got in 2007 and 2010. The results about comparison of 2007 with 2010 show in the figure. 1, figure.2 and figure.3.

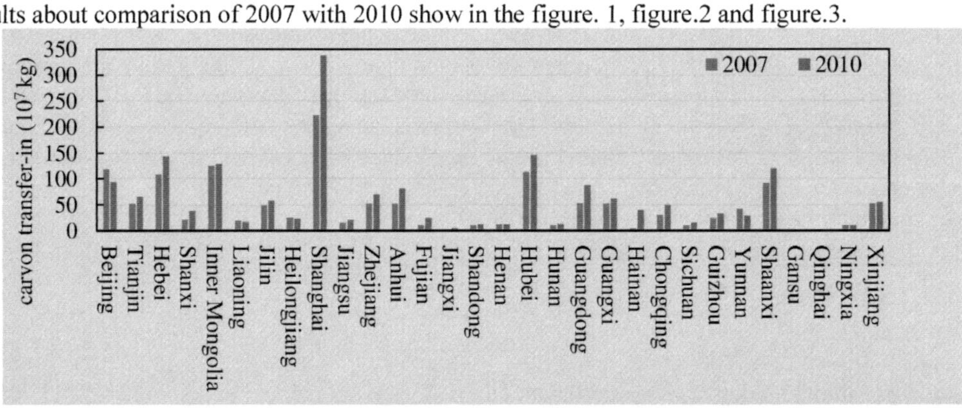

Figure 1. Comparison of provincial transportation hidden carbon transfer-in in 2007 and 2010.

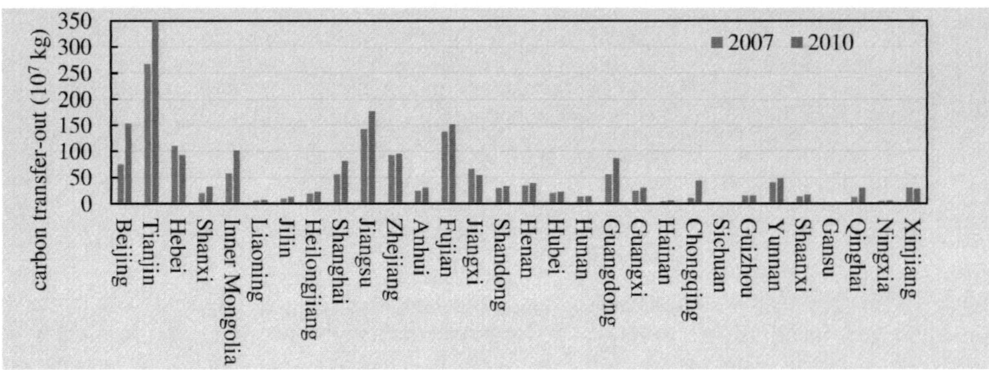

Figure 2. Comparison of provincial transportation hidden carbon transfer-out in 2007 and 2010

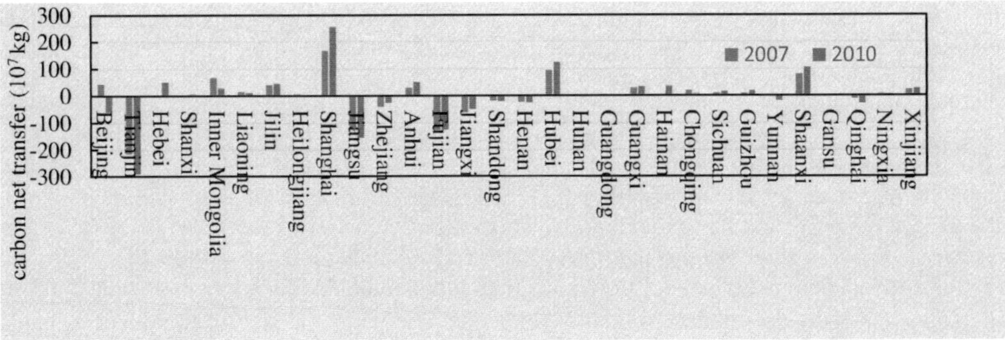

Figure 3. Comparison of provincial transportation hidden net carbon transfer in 2007 and 2010

From those we can draw the following conclusions:

(1) The following provinces had larger transportation hidden carbon transfer-in amount: Shanghai, Hebei, Hubei, Inner Mongolia, Shanxi, Beijing and Guangdong. From 2007 to 2010 years, provincial transportation hidden carbon transfer-in amount were increasing, and Shanghai had the greater growth rate among these provinces. In addition, there were also individual provinces with a downward tendency, such as Beijing.

(2) The following provinces had larger transportation hidden carbon transfer-out amount: Tianjin, Jiangsu, Beijing, Fujian, Hebei, Inner Mongolia, Zhejiang. Viewing the trend as a whole, transportation hidden carbon transfer-out amounts in provinces were increasing. Beijing and Tianjin had more obvious growth among these provinces.

(3) According to the regional structure of transportation hidden net carbon transfer, there existed obvious regional imbalance. The provinces which had more transportation hidden net carbon transfer-in were mainly located in the central and western regions. The top three regions were Shanghai, Hubei and Shanxi in both 2002 and 2007, and the last three regions were Shanxi, Yunnan and Ningxia in 2007. And in 2010 the last three regions were Heilongjiang, Guangdong and Ningxia. However, the provinces which had more transportation hidden carbon net transfer-out were mainly located in coastal areas. The top three regions were Tianjin, Jiangsu and Fujian in 2007, and the last three regions were Hainan, Hebei and Hunan. In 2010 the last three regions were Hunan, Yunnan and Shandong.e use of sections to divide the text of the paper is optional and left as a decision for the author. Where the author wishes to divide the paper into sections the formatting shown in table 2 should be used.

3.2. Transfer direction and structure analysis

By the visualization analysis tool-----ARC GIS, figure.4 displays the inter-provincial transport carbon transfer matrix.

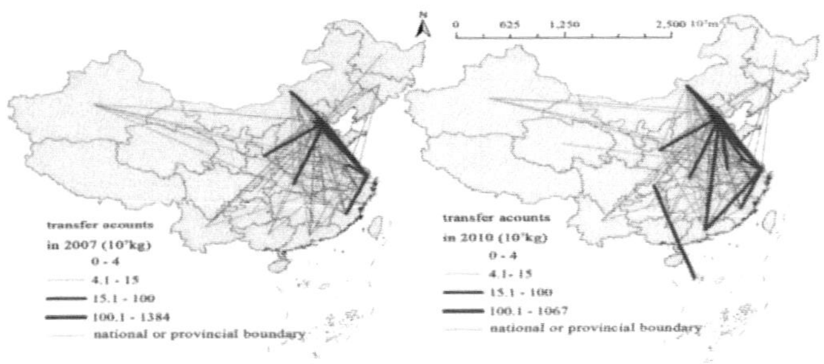

Figure 4. For the convenience of visualization, the line is the connection with the center of mass, representing the inter-provincial transportation hidden carbon transfer in 2007 and 2010.

It can be seen that transportation hidden carbon have both of the following characteristics in China:

(1)Generally speaking, the carbon transfer size showed the following characteristics: eastern regions > central regions > western regions.

(2) The transportation carbon transfer mainly occurred in the eastern-central regions, such as Beijing and Tianjin; Inner Mongolia and Tianjin; Shanghai and Beijing, Tianjin, Hebei, Inner Mongolia; Hubei and Tianjin; Shanxi and Tianjin. The size of transportation hidden carbon emissions in these provinces were more than 20 million tons, presenting the tendency of expansion.

For certain provinces, the transportation carbon emissions contain two parts, one is for developing their own economies and another one is for providing service to other provinces. From this perspective, this paper analyzes the composition of provincial transportation hidden carbon emissions in 2007 and 2010, shown in TABLE I.

As is shown in TABLE I, during 2007 and 2010, the percentage of carbon transfer-in is more than 25% in Beijing, Tianjin, Hebei and Shanxi, reflecting a quarter or more provincial transportation carbon emission is generated for improving the other provincial economics development.

However, the actual needs of transportation emissions for economic development are not only including the part generated for itself, but also including the portion transferred out to other provinces. Therefore, the actual transportation hidden carbon for provincial economic development is the sum of each column in the inter-provincial transportation carbon transfer matrix, as shown in figure 5.

Table 1. Provincial transportation hidden carbon emissions composition.

Provinces	Percentage of carbon transfer-in (%)		Provinces	Percentage of carbon transfer-in (%)	
	2007	2010		2007	2010
Beijing	30.9	19.2	Henan	3.2	2.4
Tianjin	27.5	26.3	Hubei	18.2	19.9
Hebei	27.8	32.4	Hunan	2.6	2.5
Shanxi	9.4	9.3	Guangdong	4.4	5.9
Inner Mongolia	25.9	18.3	Guangxi	15.7	14.8
Liaoning	2.4	2.0	Hainan	4.5	24.0
Jilin	22.2	21.2	Chongqing	12.2	16.1
Heilongjiang	8.1	7.7	Sichuan	2.2	2.8
Shanghai	21.0	29.8	Guizhou	13.2	12.2
Jiangsu	2.6	2.7	Yunnan	11.3	6.0
Zhejiang	10.0	11.1	Shaanxi	33.8	26.4
Anhui	23.2	29.7	Gansu	0.0	0.0
Fujian	4.2	5.7	Qinghai	0.2	0.4
Jiangxi	1.8	1.9	Ningxia	12.6	12.1
Shandong	0.9	0.9	Xinjiang	20.6	20.8

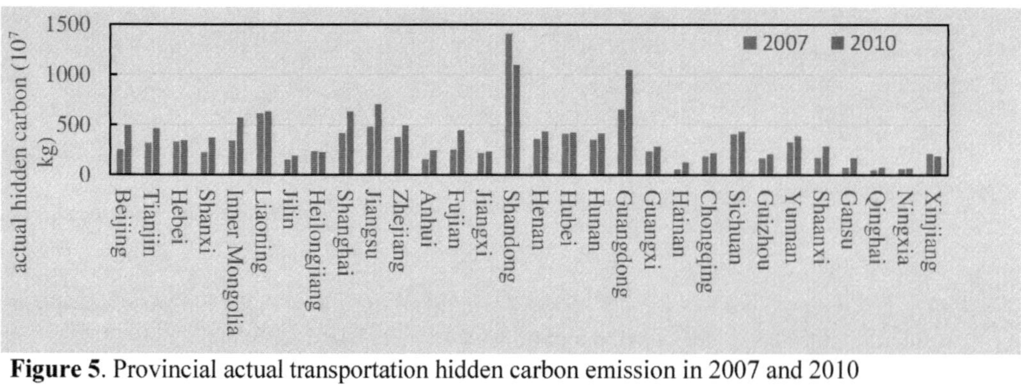

Figure 5. Provincial actual transportation hidden carbon emission in 2007 and 2010

From the regional distribution, economically developed provinces had high transportation hidden carbon emissions, like Shandong, Guangdong, Liaoning, Jiangsu and Shanghai. On the contrary, western areas were less, like Gansu, Qinghai and Ningxia. This suggests that: the more advanced the economy and the more frequently the foreign economic ties, the higher the transportation hidden carbon.

From the changing trends, the level of provincial transportation hidden carbon emission were overall on the rise during 2007 to 2010. However Shandong had been slipping, implying that the Shangdong's polices of economic structural adjustment and transportation carbon reduction had made effects.

4. Conclusions and recommendations

Based on the interregional input-output tables between China's provinces in 2007 and 2010, this paper studies the transportation hidden carbon transfer between the 30 provinces.

(1) The provinces which transportation hidden carbon transfer-in is large contained Shanghai, Hebei and Hubei. Tianjin, Jiangsu, Beijing, and other provinces had large carbon transfer-out. Among them, the provinces of net transfer-in mainly concentrated in the central-western region, such as Hubei and Shaanxi. The provinces of net transfer-out mainly concentrated in eastern coastal areas, like Tianjin, Jiangsu and Fujian.

(2) The scale of transportation carbon showed the characteristics: the eastern regions > central region > western regions. The transportation hidden carbon transfer occurred mainly in Beijing and Tianjin region, Shanghai and Beijing, Tianjin, Hebei, Inner Mongolia region, and the size of the transfer had a tendency to expand.

(3) From the structure of transportation carbon emission, 1/4 or more transportation carbon of some provinces is generated for others, such as Beijing, Tianjin, Hebei, and Shaanxi.

(4) In terms of the actual traffic hidden carbon emissions caused by economic development in provinces, Shandong, Guangdong, Liaoning, Jiangsu and other economically developed provinces' traffic hidden carbon was higher, while Gansu, Qinghai, Ningxia and other western backward areas had the low carbon traffic carbon.

Therefore, when we are designing the traffic carbon emission reduction policy in the future, we should reduce the responsibility of large transfer-in provinces appropriately, like Shanghai, Hebei, Hubei and Shaanxi, and increase responsibility of the transfer-out provinces, such as Tianjin, Jiangsu, Shandong, Guangdong and Liaoning. At the same time, for the provinces which are connected with large transfer scale, they should make the joint carbon reduction policy from the the industrial structure adjustment, such as Beijing and Tianjin, Shanghai and Beijing, Tianjin, Hebei and Inner Mongoliane.

Reference

[1] Peters G P and Hertwich E G 2008 Post-Kyoto greenhouse gas inventories: production versusconsumption *Climatic Change* **86** 51-66
[2] Lin B Q and Sun C W 2010 Evaluating carbon dioxide emissions in international trade of China *Energy Policy* **38** 613-21
[3] Muradian R, O'Connor M. and Martinez A J 2002 Embodied pollution in trade: estimating the environmental load displacement' of industrialised countries *Ecological Economics* **41** 51-67
[4] Ahmad N and Wyckoff A 2003 *Carbon dioxide emissions embodied in international trade of goods*, OECD London
[5] Sanchez J C and. Duarte R 2004 CO2 emissions embodied in international trade: evidence for Spain *Energy Policy* **32** 1999-2005
[6] Nakano S, Okamura A and Sakurai N 2009 *The measurement of CO2 embodiments in international trade: evidence from the harmonised Input-Output and bilateral trade database* OECD London

[7] Lin X P and Lu X X, 2010 International trade, pollution industry transfer and Chinese industry-ies' CO2 emissions *Economic Research Journal*. 15-26
[8] Li Y and Hewitt C N 2008 The effect of trade between China and the UK on national and global carbon dioxide emissions *Energy policy* **36** 1907-14
[9] Weber C L, Peters G P and Guan D 2008 The contribution of Chinese exports to climate change *Energy Policy* **36**, pp. 3572-7
[10] Yan L D and Yang L K China's foreign trade and climate change: a case study of CO2 emissions *Energy Policy* **38**, 350-6
[11] Xu M, Willianms E, and Allenby 2010 Assessing environmental impacts embodied in manufacturing and labor input for the China-U. S. Trade *Environmental Science & Technology* **44** 567-573
[12] Zhang X P 2009 Carbon dioxide emissions embodied in China's foreign trade *Acta Geographica Sinica* **64** 65-76
[13] Chen Y, Pan J H and Xie L H 2008 Energy embodied in goods of international trade in China: calculation and policy implications 11-25
[14] Meng L, Guo J E., Chai J and Zhang Z K 2011 China's regional CO2 emissions: characteristics, inter-regional transfer and emission reduction policies *Energy Policy* **39** 6136-44
[15] Liu H G and Fan X M 2014 CO2 emissions transfer embedded in inter-regional trade in China *Acta Ecologica Sinica*. **34** 3016-24
[16] Gao J 2013 Temporal-spatial characteristics and evolution mechanism of transport emissions PH. D. dissertation, Dept. Economics and Management, Chang'an Univ, Xi'an, China
[17] Saija S and Romano D 2002 A methodology for the estimation of road transport air emissions in urban areas of Italy *Atmospheric Environment* **36** 5377-83
[18] Hai K W, Li X F and Jun B 2011 CO2 and pollutant emission from passenger cars in China *Energy Policy* **39** 3005-11
[19] Zhang Q, Tao X M and Yang P 2012 Research on carbon emissions from metropolis urban passenger transport and countermeasures *China Population Resources and Environment* **22** 35-42
[20] Yuan C W, Rui X L and Wu D Y 2016 Carbon emission reduction pressure index for China's provincial transport sector based on geographical weighted regression model *China Journal of Highway and Transport* **29** 262-70
[21] Kwon T H 2005 Decompositon of factors determing the trend of CO2 emission from car travel in Great Britain(1970-2000) *Ecological Economics* **53** 261-75
[22] Schipper L, Saenger C and Sudardshan A 2011 Transport and carbon emission in the United States: the long view *Energies* **4** 563-81
[23] Yu J, Da Y B and Ou Y B 2015 Analysis of carbon emission changes in China's transportation industry based on LMDI decomposition method *China Journal of Highway and Transport* **28** 112-9

IOP Conf. Series: Earth and Environmental Science **51** (2017) 012008 doi:10.1088/1755-1315/51/1/012008

Removal of Suspended Solids in Anaerobically Digested Slurries of Livestock and Poultry Manure by Coagulation Using Different Dosages of Polyaluminum Chloride

P Li, C J Zhang, T K Zhao and H Zhong

Institute of Plant Nutrition and Resources, Beijing Academy of Agriculture and Forestry Sciences, No. 9 in Shuguang Garden Mid Road, Haidian District, Beijing, RR China

tkzhao@126.com

Abstract. In this study, anaerobically digested slurries of livestock and poultry manure were pretreated by coagulation–sedimentation using an inorganic polymer coagulant, polyaluminum chloride (PAC). The effect of different PAC dosages on suspended solids (SS) removal and pH in the biogas slurries was assessed to provide reference values for reducing the organic load of biogas slurry in the coagulation–sedimentation process and explore the feasibility of reducing the difficulty in subsequent utilization or processing of biogas slurry. The results showed that for the pig slurry containing approximately 5000 mg/L SS, the removal rate of SS reached up to 81.6% with the coagulant dosage of 0.28 g/L PAC. For the chicken slurry containing approximately 2600 mg/L SS, the removal rate of SS was 30.2% with the coagulant dosage of 0.33 g/L PAC. The removal rate of SS in both slurries of livestock and poultry manure exhibited a downward trend with high PAC dosage. Therefore, there is a need to control the PAC dosage in practical use. The pH changed little in the two types of biogas slurries after treatment with different PAC dosages and both were in line with the standard values specified in the "Standards for Irrigation Water Quality".

1. Introduction

The large-scale intensive livestock and poultry industry has developed rapidly in recent years in China. The gross output value of the livestock industry was 16.27 billion yuan in Beijing in 2011, an increase of 63.8% over 2001. This development has been accompanied by the production of a large amount of animal manure and sewage, among other wastes. Livestock and poultry wastes have become a major source of nonpoint source pollution in rural areas in China [1]. According to China's first national census of pollution sources [2], the national production of livestock and poultry wastewater was 1.321 billion m3, 57.3% of which was processed and used; the production of livestock and poultry wastewater in Beijing was 4.7774 million m3, only 56.0% of which was processed and used—lower than the national level. The pollutant emissions from agricultural sources have a significant impact on the aquatic environment. Specifically, the chemical oxygen demand (COD) emissions from agriculture reached 13.2409 million t; that is, 43.7% of the total COD emissions across the country. The COD emissions from livestock and poultry industry accounted for 95.8% of the total emissions from agricultural sources, leading to prominent pollution problems in the livestock and poultry industry. Anaerobic fermentation processing technology (i.e., biogas production) plays a major role in the disposal of large amounts of livestock and poultry manure [3], and it has become an important measure of waste treatment processes in livestock and poultry farms [4]. In China, the number of biogas users reached 3.0×107 in 2009, ranking first in the world [5]. Additionally, there were more than 4700 large- and medium-sized biogas plants on large-scale farms in 2010, which produced 3 billion t of biogas slurry and residue [6].

Anaerobically digested slurry produced after biogas production by anaerobic fermentation, also known as biogas slurry, contains abundant nitrogen, phosphorus, and potassium, among other major elements; meanwhile, biogas slurry contains large amounts of trace elements (e.g., iron, zinc, and copper), organic matter, many amino acids and vitamins, as well as various beneficial bacteria [7-10]' and it can be returned to farmland as fertilizer [11]. Despite the particular benefits obtained from the use of biogas slurry in agriculture worldwide [12-13], the direct discharge of biogas slurry with high pollutant load into farmland can cause detrimental effects to the environment [14-16]. COD and heavy metals are major obstacles that affect the emission and utilization of biogas slurry [12, 17, 18]. Moreover, the application of biogas slurry is limited by arable land resources, storage and transportation costs [19], and crop demand, among various factors [20]. Thus, biogas slurry needs to be effectively processed to ensure the convenience and environmental safety of resource reuse.

Anaerobically digested manure slurry has significantly high concentrations of suspended solids (SS) and COD even after long-term standing and precipitation [21]. Once high-COD effluent is discharged into farmland, the organic matter contained in it will be decomposed; the intermediates can cause crop yield losses, while methane, a greenhouse gas, can also be produced, posing a threat to the atmosphere. The COD of biogas slurry is mainly composed of insoluble organic matter [21] that is unfavorable for subsequent use and processing. Removal of high concentrations of SS can markedly reduce the organic pollution load of biogas slurry, thereby facilitating the subsequent processing. The most simple, effective, and relatively low-cost method of SS removal is coagulation–sedimentation [22]. An important factor that affects the coagulation effect is the dosage of coagulant and coagulant aids [23]. There were no researches for assessing the effect of SS removal on anaerobically digested slurries of pig and chicken manure. In the present study, anaerobically digested slurries of pig and chicken manure were used to assess the effect of different dosages of cheap coagulation agent (i.e., polyaluminum chloride, PAC) for removal of SS. The results will provide a reference for reducing the organic load in biogas slurries of livestock and poultry manure.

2. Materials and methods

2.1. Materials

Anaerobically digested slurry of pig manure (hereinafter referred to as pig slurry) was obtained from the waste effluent tank of a biogas plant in a medium-sized pig farm in Yanqing District, Beijing, China. Anaerobically digested slurry of chicken manure (hereinafter referred to as chicken slurry) was obtained from the waste effluent tank of a biogas station in Yanqing District, Beijing. To prevent degradation at room temperature, the test samples were immediately stored at 4°C upon arrival at the laboratory. The test slurries were passed through a 60-mesh standard sieve to remove coarse impurities before use. The SS content and pH of the test slurries are shown in Table 1. The coagulant was commercially available PAC (an inorganic macromolecular polymer of aluminum salt).

Table 1. Pollutant content in biogas slurries tested in the study

	Suspended solid content (mg/L)	pH
Pig slurry	5080	7.84
Chicken slurry	2693	7.85

2.2. Methods

2.2.1. Experimental design. In this experiment, a 1000-mL PVC beaker was used as the reaction vessel and a jar test mixer was used for agitation. First, a preliminary experiment was conducted to determine the coagulant dosage for the formation of alumen ustum. Briefly, 400 mL of sieved slurry was added into the beaker and placed on the mixer platform. The mixer was agitated slowly at 60 rpm. When the agitation speed stabilized, the coagulant was added at 0.02 g PAC each time until the appearance of alumen ustum during agitation. The cumulative amount of PAC added was recorded as the minimum coagulant dosage for the formation of alumen ustum, xi (g/L). The experiment was performed three times and the mean value of x1, x2, and x3 was calculated.

Based on the minimum coagulant dosage for the formation of alumen ustum, x (g/L), we chose x/3 as the coagulant dosage for Beaker #1, 2x as the coagulant dosage for Beaker #6, and the coagulant dosage was successively and equally increased for Beakers #2–5. The coagulant was added into Beakers #1–6 simultaneously and the rates of the coagulant dosage are shown in Table 2. Next, 800 mL of sieved slurry was added into the reaction vessel on the mixer platform. The mixer program was run as follows: rapid agitation at 250 rpm for 1 min with coagulant addition, intermediate agitation at 120 rpm for 10 min, slow agitation at 30 rpm for 10 min, and then agitation was stopped. The water sample was allowed to stand and precipitate for 10 min. The supernatant was collected from the reaction vessel using a 50 mL syringe and used to measure the SS content and pH.

Table 2. Treatments with different coagulant dosages

Beaker No.	1	2	3	4	5	6
Dosage (g/L)	x/3	2x/3	x	4x/3	5x/3	2x

2.2.2. Analytical methods. The SS content of slurry samples was determined by the gravimetric method (GB11901-89). The samples were filtered through a 0.45-μm pore size membrane filter. The filter membrane was dried at 105°C to constant weight and then weighed. The slurry pH was measured using the glass electrode method.

The experimental data were statistically analyzed using SPSS Statistics 19.0 (SPSS Inc., Chicago, IL, USA). The results were plotted using Origin 8.5 (OriginLab Corp., Northampton, MA, USA).

3. Results and discussion

3.1. Effect of coagulation pretreatment on SS in pig slurry

According to the results of the preliminary experiment, 0.83 g/L PAC was the minimum coagulant dosage to form alumen ustum in the pig slurry during coagulation treatment. Therefore, the PAC dosages for the pretreatment of pig slurry were 0.28, 0.55, 0.83, 1.11, 1.39, and 1.66 g/L.

With the PAC dosage in the range of 0.28–0.83 g/L, the SS content of pig slurry pretreated by coagulation–sedimentation was lower than 1000 mg/L and the removal rate of SS remained higher than 80% (Figure 1). With a continued increase in PAC dosage, the SS content of pretreated pig slurry showed an upward trend, while the removal rate of SS dropped to below 80%, varying from 75.9% to 66.4%. With different PAC dosages, the pH of pretreated pig slurry varied between 7.85 and 8.03 (Figure 2), which was slightly but not significantly increased compared with the original slurry; all pH values were in line with the standard values (5.5 to 8.5) specified in the "Standards for Irrigation Water Quality". With an increase in PAC dosage, the slurry pH showed a slight increase followed by a progressive decrease. The highest pH value was observed with the coagulant dosage of 0.55 g/L PAC.

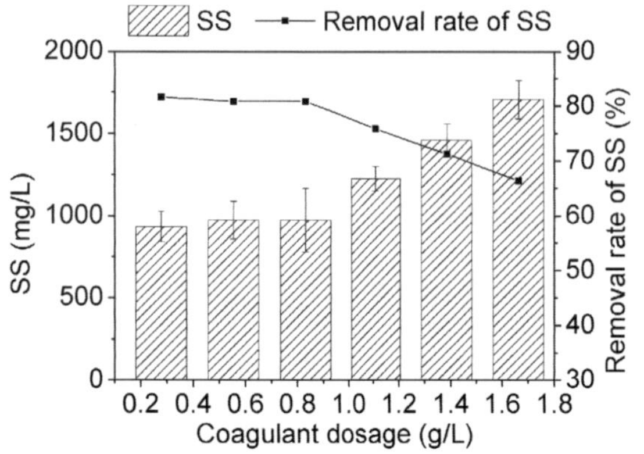

Figure 1. The content and removal rate of suspended solids (SS) in pig slurry pretreated by coagulation–sedimentation with different coagulant (polyaluminum chloride) dosages.

In the coagulation–sedimentation process, we can preliminarily and intuitively determine the result of coagulation treatment through the formation, size, and sedimentation rate of alumen ustum. There are multiple influencing factors: in addition to the hydraulic conditions and raw water quality, the type and dosage of coagulants strongly influence the formation, size, and sedimentation rate of alumen ustum. If the coagulant dosage is insufficient, it is difficult to form alumen ustum and the effluent is turbid; if the coagulant dosage is too high, the size of alumen ustum decreases and the effluent is also turbid. When the coagulant dosage is within a suitable range, the best results can be obtained with regard to the formation, size, and sedimentation rate of alumen ustum [24]. Han et al. [25] treated the biogas slurry of cattle manure (hereinafter referred to as cattle slurry) with PAC. It was found that with an increase in PAC dosage, the removal rate of cattle slurry turbidity first increased and then decreased, with the highest removal rate obtained with the dosage of 320 mg/L PAC. In the current study, the highest removal rate of SS was obtained in the pig slurry with the dosage of 0.28–0.55 g/L PAC, in agreement with previous results[25].

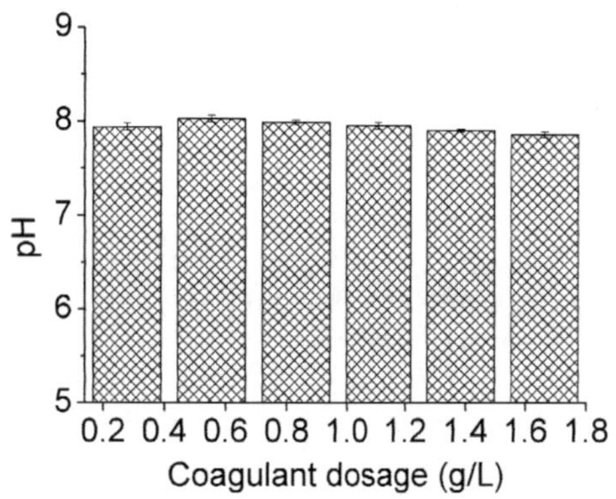

Figure 2. The pH in pig slurry pretreated by coagulation–sedimentation with different coagulant (polyaluminum chloride) dosages.

3.2. Effect of coagulation pretreatment on SS in chicken slurry

According to the results of the preliminary experiment, approximately 0.25 g/L PAC was the minimum coagulant dosage to form alumen ustum in the chicken slurry treated by coagulation–sedimentation. Therefore, the PAC dosages for pretreatment of chicken slurry were 0.08, 0.17, 0.25, 0.33, 0.42, and 0.50 g/L.

With an increase in PAC dosage, the SS content of chicken slurry pretreated by coagulation–sedimentation showed a slight decrease followed by an increase (Figure 3). With the PAC dosage in the range of 0.33–0.42 g/L, the SS content of pretreated chicken slurry was lower than 1900 mg/L and the removal rate of SS remained higher than 30%. With a continued increase in PAC dosage, the SS content of pretreated chicken slurry markedly decreased, while its removal rate decreased to 10.9%. The slurry pH varied between 7.89 and 7.91 after pretreatment with different PAC dosages (Figure 4).

The pH value increased slightly, but not significantly, compared with the original slurry, and all pH values were in line with the standard values specified in the "Standards for Irrigation Water Quality". The slurry pH was almost constant with the increase in PAC dosage, and relatively high pH was detected with the PAC dosages of 0.08 and 0.25 g/L.

Han et al. reported that when cattle slurry was treated with PAC, the highest removal rate of slurry turbidity was obtained with 320 mg/L PAC; however, the removal rate of slurry turbidity declined when the PAC dosage reached 640 mg/L [25]. In the present study, the highest removal rate of SS was obtained in the chicken slurry with the PAC dosage of 0.33 g/L; the removal rate of SS significantly decreased when the dosage reached 0.50 g/L. The trend of SS removal rate in the chicken slurry was consistent with that reported by Han et al. in cattle slurry [25]. The inorganic macromolecular polymer of aluminum can provide a large number of complex ions, which strongly adsorb colloidal particles and also aggregate colloids through adsorption bridging. The simultaneous chemical and physical changes neutralize the surface charge of the colloidal particles and reduce the zeta potential, resulting in unstable micelles and causing floc formation and sedimentation [26]. With insufficient PAC dosage, impurities in the water do not receive adequate reverse charges for destabilization or coagulation to occur. With excessive PAC dosage, the opposite charge on the colloid surface allows the slightly destabilized colloids to regain stability. According to a domestic study on the removal of organic compounds in chicken slurry by coagulation [27], the pollutant content was relatively low in the chicken slurry tested in the present study. Thus, a comparably good rate of SS removal was obtained with low coagulant dosage of PAC. With a further increase in PAC dosage, the particles in the biogas slurry are surrounded by too much coagulant and thus lose the opportunity to combine with other particles, making coagulation unlikely to occur and thereby affecting the removal of SS [27].

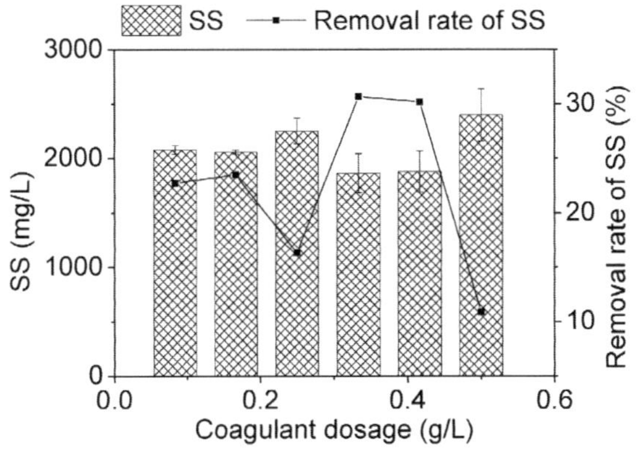

Figure 3. The content and removal rate of suspended solids (SS) in chicken slurry pretreated by coagulation–sedimentation with different coagulant (polyaluminum chloride) dosages.

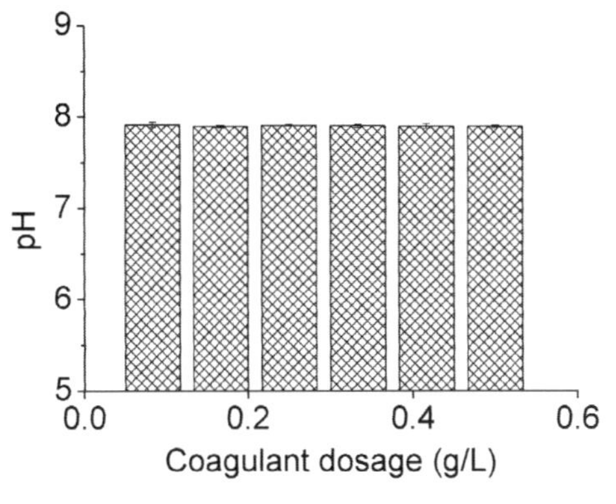

Figure 4. The pH in chicken slurry pretreated by coagulation–sedimentation with different coagulant (polyaluminum chloride) dosages.

4. Conclusions

With the coagulant dosage in the range of 0.28–0.83 g/L PAC, the removal rate of SS reached more than 80% in the anaerobically digested pig slurry containing approximately 5000 mg/L SS; the highest removal rate was obtained with the PAC dosage of 0.28 g/L. With the coagulant dosage in the range of 0.33–0.42 g/L PAC, the removal rate of SS was higher than 30% in the anaerobically digested chicken slurry containing approximately 2600 mg/L SS; the highest removal rate was obtained with the PAC dosage of 0.33 g/L. High PAC dosage resulted in a decrease in the removal rate of SS in both slurries of livestock and poultry manure, making it necessary to control the PAC dosage in practical use. The pH changed little in the two types of biogas slurries after treatment with different PAC dosages and both complied with the standard values specified in the "Standards for Irrigation Water Quality".

Acknowledgments
This work was supported by the National Science & Technology Pillar Program during the 12th Five-Year Plan period (2012BAD15B02) and the Technological Innovation Team Development of BAAFS (JNKYT201605).

References
[1] Zhang J, Zhang M X, Shan S D, et al 2009 *Journal of Agro-Environment Science*. 28(10): 2005-2009
[2] Compilation committee of the data of the first national census of pollution sources 2011 *Data collection of the first national census of pollution sources* vol 4 (*Technical report of pollution sources census* vol 2) (Beijing: China Environmental Press) pp 42-47
[3] Li S and Deng L W 2008 *Swine Industry Science*. (1): 70-72
[4] Zhu J, Qiu Y Q, Zhang C P 2010 *Journal of Safety and Environment*. 10(5): 24-28
[5] Zhang J F, Yuan H R, Zou D X, et al 2012 *Journal of Anhui Agri. Sci*. 40(19): 10246-10250
[6] Xie J J, Xu F H, Qi J Y, et al 2012 *China biogas*. 30(5): 14-18
[7] Zeng G K, Xie J, Yin F 2005 *Renewable Energy*. (1): 38-40
[8] Ren N Q, Wang A J, Ma F 2005 *Microbial physiological ecology of acid producing biogs microorganism* (Beijing: Science Press)
[9] Tani M, Sakamoto N, Kishimoto T, et al 2006 *International Congress Series*. 1293: 331-334
[10] Marcato C E, Pinelli E, Pouech P, et al 2008 *Bioresource Technology*. 99: 2340-2348
[11] Matsunaka T 2006 *International Congress Series*. 293: 242-252
[12] Liu X L, Liu J W, Liu B 2012 *Journal of Anhui Agri. Sci*. 40(2): 968-971
[13] Xu Q X, Guan X F, Qian L, et al. 2015 *Environmental Engineering*. 33(1): 72-76
[14] Noykova N, Müller T G, Gyllenberg M, et al. 2002 *Biotechnol Bioeng*. 78(1): 89-103
[15] Mahmoud N, Zeeman G, Gijzen H, et al. 2004 *Water Research*. 38(9): 2348-2358
[16] Myint M, Nirmalakhandan N, Speece R E. 2007 *Water Research*. 41(2): 323-332
[17] Ge X, Li B Q, Ding Y Q, et al. 2012 *Journal of Anhui Agri. Sci*. 40(30): 14897-14898, 15058
[18] Christos M, Christos S, Nikolaos R, et al. 2012 *Journal of Agricultural Science and Technology*. A2: 149-154
[19] Waeger F, Delhaye T, Fuchs W 2010 *Separation and Purification Technology*. 73: 271-278
[20] Bai X F, Li Z F, Cheng S K, et al. 2014 *Envionmetal Engineering*. 32(6): 153-156
[21] Tang L Q, Zhu H G, Chen J, et al. 2010 *China Biogas*. 28(6): 7-12
[22] Xiong H J and Liu S R 2011 *Environmental Protection of Oil & Gas Fields*. 21(2): 30-33
[23] Ahmad A L, Ismall S, Bhatia S 2005 *Environmental Science and Technology*. 39(8): 2828-2834
[24] Zhou L Y, Wang S, Ni J 2008 *Shanghai Water*. (4): 14-17
[25] Han M, Feng C H, Liu K F, et al. 2014 *Tianjin Agricultural Sciences*. 20(9): 96-101
[26] Zhu H G, Wang D Y 2012 *Transactions of the Chinese Society for Agricultural Machinery*. 43(4): 93-99, 118
[27] Zhao Y 2006 *Research on removal of phosphorus and turbidity in municipal wastewater by PACl*, (Wuhan: Central China Normal University)

Research of Isolation and Degradation Conditions of Petroleum Degrading Marine

Guo Fangrui

School of Marine Science and Technology, Harbin Institute of Technology at Weihai, Harbin, China

guofangrui_2016@163.com

Abstract. A novel petroleum-degrading microbial strain was isolated from sediment samples in estuary of Bohai Sea estuary beaches. The strain was primarily identified as Alcanivorax sp. and named Alcanivorax sp. H34. Effect of PH values, temperature, nitrogen and phosphorus concentrations on degradation of H34 were investigated. The paraffinic components average degradation rate of H34 ungrowth cells under optimized conditions was studied. The results showed that the optimal growth conditions of H34 are were temperature of 30°C, initial PH of 7.0, nitrogen concentration of 3g/L, phosphorus concentration of 3g/L, and paraffinic components average degradation rates of H34 ungrowth cells was 41.6%, while total degradation rate was 45.5%.

1. Introduction

With the improvement of the petroleum extraction technology and development of petroleum industry, the maritime transport increased year by year. At the same time, the marine petroleum spills happens every year which results in large amounts of petroleum leaked into the sea and great harm to human life. Compared with chemical and physical methods, biological methods is an important way to completely eliminate the spilled petroleum by the microbial degradation. It can not only help us to understand the self-healing capacity of the environment, but also provide bioremediation with strain resources. Due to the hydrophobicity of the petroleum hydrocarbon, the microbial degradation of petroleum hydrocarbons is low. Thus, it is significant to isolate and identify degrading strains from the polluted seawater, and improve the microbial degradation by optimize their growth conditions.

There has been a growing environmental and research concern toward petroleum spills due to their wide distribution, persistence and toxicity to humanbeings and other biota. To effectively solve this problem, many methods are proposed. The result is demonstrate clearly that the bacterial consortium MPD-M was effective for treating hydrocarbons in water with salinity varying from 0 to 180g/L [1]. The efficiency of Bacillus subtilis DM-04 and Pseudomonas aeruginosa M and NM strains isolated from a petroleum contaminated soil sample from North-East India was compared for the biodegradation of crude petroleum-oil hydrocarbons in soil and shake flask study[2]. Two types of Indian crude oil (Bombay High and Gujarat) were tested for their biodegradability by Acinetubacter cakoaceticus and Alcnligenes odornns [3]. A mixed consortium was prepared with 15 bacteria isolated by enrichment technique from the sample collected from an oil contaminated site, and the degradation efficiency of the isolates in consortium was checked with 2% crude oil by shake flask transformation in mineral salt medium, at 37°C for 24 days [4]. The phenanthrene degrading novel bacterium strain USTB-RU was isolated from petroleum contami-nated soil in Dagan oilfield, southeast of Tianjin, northeast China [5]. Isolation and identification of biosurfactant producing bacteria were assessed, and the potential application of these bacteria in petroleum industry was investigated [6].

In this paper, we name the petroleum-degrading microbial strain which is isolated tidal flat sediment in estuary of Bohai as Alcanivorax sp. H34, since it is primarily identified as Alcanivorax sp. Then, the effect of PH values, temperature, nitrogen and phosphorus concentrations on degradation of Alcanivorax sp. H34 is investigated. In addition, we also study the content and degradation rate of Alcanivorax sp. H34 non-growing cells alkane degradation

2. Materials And Methods

The first paragraph after a heading is not indented (Bodytext style).

Other paragraphs are indented (BodytextIndented style).

2.1. Sampling

Sediment samples were collected from the Bohai Sea estuary beaches, and the oil samples were taken from the Dagang Oilfield.

2.2. Medium

2.2.1. Enrichment medium. Adjust pH of the medium (composition: NH_4NO_3 50g, KH_2PO_4 30g, K_2HPO_4 9g, $MgSO_4$-$7H_20$ 0.4g) to 7.0. Sterilize by autoclaving at 121°C for 15 minutes. Add diesel 5mL. Sterilize by autoclaving at 105°C for 15 minutes.

2.2.2. Isolation. Add the following to 300mL H_2O; Tryptone 0.5g; Yeast extract 0.5g; Beef extract 2g; Sodium acetate 0.2g; Agar 15g. Adjust pH to 7.0. Sterilize by autoclaving at 121°C for 15 minutes. Add 700mL seawater, and Sterilize by microporous membrane with 14um.

2.2.3. Re-screening medium. Sterilize 50mL seawater by microporous membrane with 14um. Add NH_4NO_3 50g, KH_2PO_4 30g, K_2HPO_4 9g ,$MgSO_4$-$7H_20$ 0.4g. Sterilize by autoclaving at 121°C for 15 minutes. Add diesel 5mL. Sterilize by autoclaving at 105°C for 15 minutes.

2.3. Enrichment, Isolation and Selection of Crude-oil Degrading Bacteria

The 30/300mL Enrichment medium was added to Erlenmeyer flasks and the flasks were incubated for 7 days at 30°C on rotary shaker with 120 rpm. Then the medium was diluted with 1 times, 100 times, 1,000 times with saline, and separated by the separation medium with streak plate method. After that, the solution was cultured by inverted until the colonies outgrow. At last, take the single colony, and isolated repeatedly using the streak plate method. Inoculated into 30/300mL re-screeningm medium. Incubated for 7 days at 30°C on rotary shaker with 120 rpm.

2.4. Alcanivorax sp.
H34 Non-growing Cells Degrade Petroleum Hydrocarbons

2.4.1. Strains cultured. The strain Alcanivorax sp. H34 was inoculated to tubes containing 5mL FM medium. Activated for 24 hours under the condition of 30°C and 120rpm. Two active strains were inoculated to two 2216E mediums (30mL and 300mL) with 5% inoculum. Incubated for 48 hours at 30°C and 120 rpm.

2.4.2. Collected cells, inoculated, degradation. Take 30mL H34 broth, centrifugate at 14000 rpm and 4°C for 15min, and discard the supernatant. Transfer the centrifugal thallus to degradation of petroleum hydrocarbon using 3mL sterilization natural seawater. Degradate 7 days at 3°C and 20 rpm.

2.4.3. Diesel group for Comparison. Degradation of petroleum hydrocarbon was sterilized by microporous membrane. Add to the flask, and degraded for 7-8 days at 30°C, 120 rpm.

3. Experimental Results and Analysis

To research the optimal growth conditions of H34, temperature, initial PH value, nitrogen and phosphorus concentrations were studied. Put H34 to 5ml FM medium under condition of 30°C, 120 rprn, activated for 24 hours. Then, put 5% inoculation amount to hydrocarbon-degrading matrix, 5mL tubes, 120 rpm and measured OD_{600} every 24 hours.

3.1. Influence of Temperature on Alcanivorax sp. H34

To research the influence of temperature on Alcanivorax sp. H34, set the temperature 15°C and 30°C, the initial PH value 7.0, the NH_4NO_3 concentration 3 g/L, K_2HPO_4 3 g/L, KH_2PO_4 1 g/L. the result is shown in Fig.1.

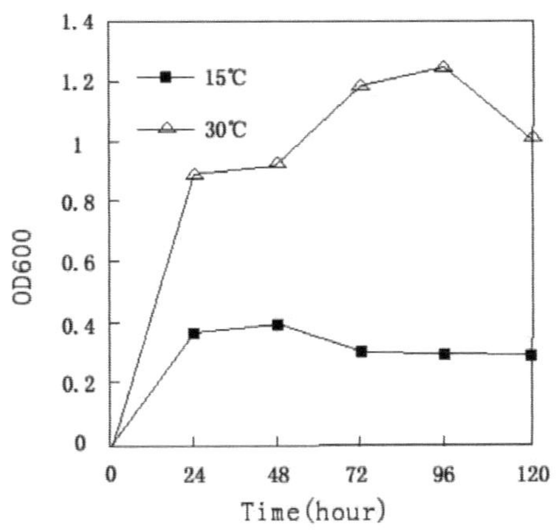

Figure 1. Influence of temperature on Alcanivorax sp. H34 consumption of alkanes

From Fig.1 we can see that during the growth of Alcanivorax sp. H34, the strain can grow at two different temperatures, the growth reaches the maximum at 96 h, and the optimum growth temperature of Alcanivorax sp. H34 is 30 °C.

3.2. Influence of Initial PH on Alcanivorax sp. H34

The optimal initial PH value of Alcanivorax sp. H34 consumption of alkanes is researched under the condition of temperature 30°C, the NH_4NO_3 concentration 3g/L, K_2HPO_4 3g/L, KH_2PO_4 1 g/L. The initial PH values were set 6, 7, 8 and 9. Influence of pH on Alcanivorax sp. H34 consumption of alkanes is shown in Fig. 2.

Fig. 2 shows that during the growth of Alcanivorax sp. H34, the strain can grow at four different initial PH values, the growth reaches the maximum at 96 h, and the optimum growth initial PH value of Alcanivorax sp. H34 is 7.0. PH value affects the decomposition of petroleum hydrocarbons, and also inhibits the growth and metabolism of strains.

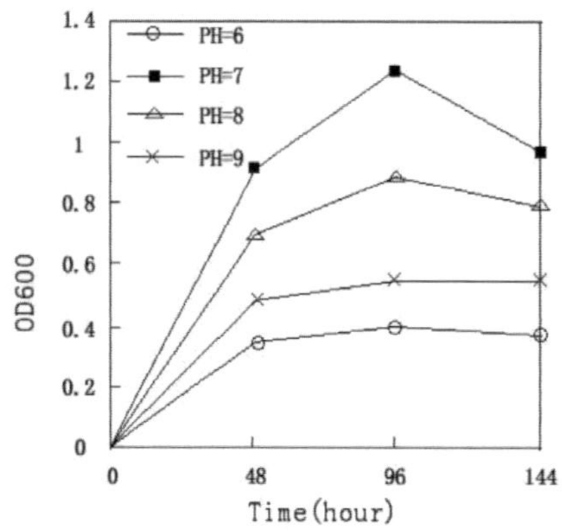

Figure 2. Influence of pH on Alcanivorax sp. H34 consumption of alkanes

3.3. Influence of Nitrogen Concentration on Alcanivorax sp. H34

The optimal nitrogen concentration of Alcanivorax sp. H34 consumption of alkanes is researched under the condition of temperature 30°C, initial PH value 7.0 , K_2HPO_4 3g/L, KH_2PO_4 1g/L. The nitrogen concentrations are set 1g/L, 2g/L, 3g/L, 4g/L and 5g/L. Influence of nitrogen concentrations on Alcanivorax sp. H34 consumption of alkanes is shown in Fig. 3.

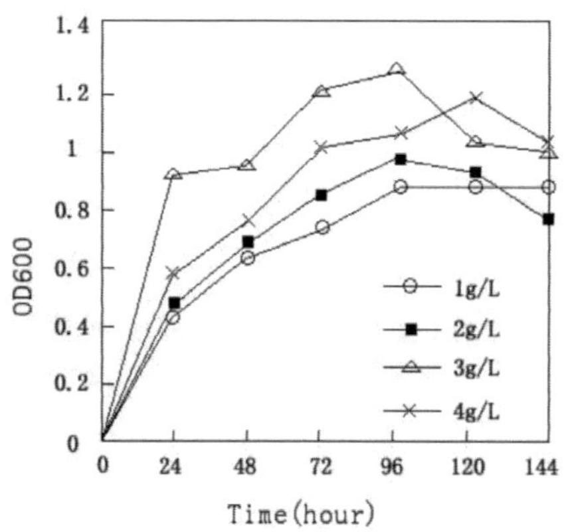

Figure 3. Influence of nitrogen concentrations on Alcanivorax sp. H34 consumption of alkanes

From Fig.3 we can see that during the growth of Alcanivorax sp. H34, the strain can grow at four different nitrogen concentrations. When the nitrogen concentration of Alcanivorax sp. H34 is 3g/L, the strains grow best and the growth reaches the maximum when 96h.

3.4. Influence of Phosphorus Concentration on Alcanivorax sp. H34

The optimal phosphorus concentration of Alcanivorax sp. H34 consumption of alkanes is researched under the condition of temperature 30°C, initial PH value 7.0, the nitrogen concentration set 3g/L.

Phosphorus concentrations are 1g/L, 2g/L, 3g/L, 4g/L and 5g/L. Influence of phosphorus concentrations on Alcanivorax sp. H34 consumption of alkanes is shown in Fig. 4.

Fig.4 shows that during the growth of Alcanivorax sp. H34, the strain can grow at four different phosphorus concentrations. When the phosphorus concentration of Alcanivorax sp. H34 is 3g/L, the strains grow best and when 144h, the growth reaches the maximum.

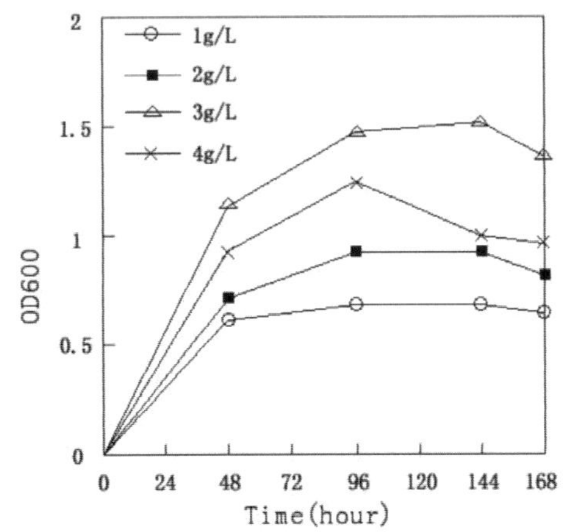

Figure 4. Influence of phosphorus concentrations on Alcanivorax sp. H34 consumption of alkanes

3.5. Alcanivorax sp. H34 Non-growing Cells Alkane Degradation

To research the Alcanivorax sp. H34 non-growing cells alkane degradation, experiments are carried by the method shown in section 2.4 and the results are shown in Fig. 5.

From Fig.5 we can see that Alcanivorax sp. H34 non-growing cells can degrade moderate carbon chain (CIQ - C24). The paraffinic components average degradation rates of H34 ungrowth cells is 41.6%, total degradation rate is 45.5%. The degradation rate slightly varies for different components. With the increasing of carbon number, the degradation rate decreases. The degradation rate of C10-C18 is high, 55.7% the highest and 21.6% the lowest.

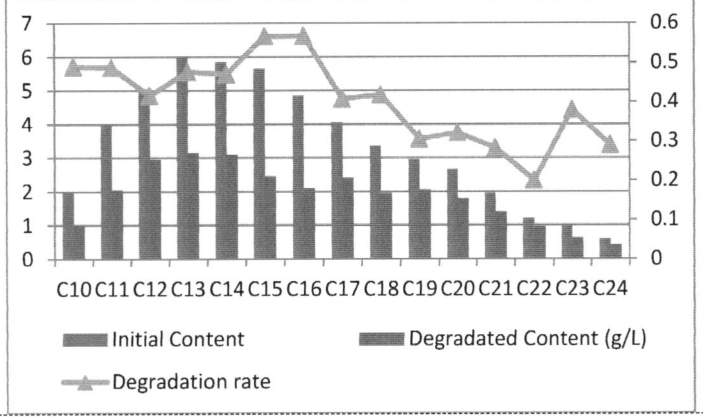

Figure 5. Content and degradation rate of Alcanivorax sp. H34 non-growing cells alkane degradation

4. CONCLUSIONS

During the growth of Alcanivorax sp. H34, the optimum growth temperature is 30 °C, the optimum growth initial PH value is 7.0 and the growth reaches the maximum at 96 h. When the nitrogen concentration of Alcanivorax sp. H34 is 3g/L, the strains grow best and when 96h, the growth reaches the maximum. When the phosphorus concentration of Alcanivorax sp. H34 is 3g/L, the strains grow best and when 144h, the growth reaches the maximum.

Alcanivorax sp. H34 non-growing cells can degrade moderate carbon chain (CIQ - C24). The paraffinic components average degradation rates of H34 ungrowth cells is 41.6%, total degradation rate is 45.5%. The degradation rate slightly varies for different components. With the increasing of carbon number, the degradation rate decreases. The degradation rate of C10-C18 is high, 55.7% the highest and 21.6% the lowest.

References

[1] M. P. Díaz, K. G. Boyd, S. J. W. Grigson, et al., "Biodegradation of crude oil across a wide range of salinities by an extremely halotolerant bacterial consortium MPD-M, immobilized onto polypropylene fibers," Biotechnology and bioengineering, vol. 79, no. 2, pp. 145-153, 2002.

[2] K. Das and A. K. Mukherjee, "Crude petroleum-oil biodegradation efficiency of Bacillus subtilis and Pseudomonas aeruginosa strains isolated from a petroleum-oil contaminated soil from North-East India," Bioresource technology, vol. 98, no. 7, pp. 1339-1345,2007.

[3] B. Lal and S. Khanna, "Degradation of crude oil by Acinetobacter calcoaceticus and Alcaligenes odorans," Journal of applied bacteriology, vol. 81, no. 4, pp. 355-362, 1996.

[4] Z. Malik and S. Ahmed, "Degradation of petroleum hydrocarbons by oil field isolated bacterial consortium," African Journal of Biotechnology, vol. 11, no. 3, pp. 650-658, 2014,

[5] K. Masakorala, J. Yao, M. Cai, et al., "Isolation and characterization of a novel phenanthrene (PHE) degrading strain Psuedomonas sp. USTB-RU from petroleum contaminated soil," Journal of hazardous materials, vo. 263, pp. 493-500, 2013.

[6] A. Tabatabaee, M. M. Assadi, A. A. Noohi, et al., "Isolation of biosurfactant producing bacteria from oil reservoirs", 2005.

The Wireless Sensor Network (WSN) Based Coal Ash Impoundments Safety Monitoring System

E J Sun[1, 2]**, A Nieto** [3, 4] **and X K Zhang**[1]

[1] China Academy of Safety Science and Technology, Beijing, 100012, CHINA
[2] Mining and Mineral Department, Virginia Tech, VA 24060-7550, USA
[3] Laureate Education 650 S. Exeter Street Baltimore, Maryland 21202-4382, USA
[4] 122 Hosler Bldg., Engineering, Penn State University, Penn State, 16802, USA

enjisun@gmail.com

Abstract. Coal ash impoundments are inevitable production of the coal-fired power plants. All coal ash impoundments in North Carolina USA that tested for groundwater contamination are leaking toxic heavy metals and other pollutants. Coal ash impoundments are toxic sources of dangerous pollutants that pose a danger to human and environmental health if the toxins spread to adjacent surface waters and drinking water wells. Coal ash impoundments failures accidents resulted in serious water contamination along with toxic heavy metals. To improve the design and stability of coal ash impoundments, the Development of a Coal Ash Impoundment Safety Monitoring System (CAISM) was proposed based on the implementation of a wireless sensor network (WSN) with the ability to monitor the stability of coal ash impoundments, water level, and saturation levels on-demand and remotely. The monitoring system based on a robust Ad-hoc network could be adapted to different safety conditions.

1. Introduction

Coal-fired power plants is the world's largest coal consumer industry. Coal consumption accounts for almost 70% of China's primary energy consumption. The proportion of coal-fired power plants is more than 50% of national coal consumption. [1] The amount of coal ash emissions from coal-fired power plants have increased 2.5 times in the past eight years. The production of coal ash has reached 375 million tons, which can reach 424 million cubic meters. It has become the largest single source of industrial solid waste emissions in China. [2] Coal ash is inevitable product of coal-fired power plants and its amount produced by power plants increases along with the expansion of installed capacity of power generators. [3] It is the toxic waste formed from burning coal in power plants to make electricity. The second largest industrial waste stream in the U.S. According to the American Coal Ash Association's Coal Combustion Product Production & Use Survey Report, nearly 110 million tons of coal ash was generated in 2012. A total of 1.25 billion tons of coal ash is stored in China with an annual emission of 1.5 million tons. [4] Hundreds of contaminated sites and spills have been documented among the 1,400+ coal ash waste dumps across the U.S. [5] Many coal ash dumps lack basic safety features and regular inspections, leaving communities at risk of large scale disasters like those in Kingston, Tennessee and North Carolina. [6] Coal ash impoundments are inevitable production of the coal-fired power plants. It contains a range of toxic constituents that are known to leach, leak, or spill out of coal ash disposal sites and adversely affect human and environmental health.

[7] Two dozen homes were destroyed or damaged by the 2008 coal ash spill in Kingston, Tennessee as shown in Figure 1.

Figure 1. Two dozen homes were destroyed or damaged by the 2008 coal ash spill in Kingston, Tennessee.

All coal ash impoundments in NC, USA that have been tested for groundwater contamination are leaking toxic heavy metals and other pollutants. Coal ash impoundments are toxic sources of dangerous pollutants that pose a danger to human and environmental health if the toxins spread to adjacent surface waters and drinking water wells. Several coal ash impoundments failures produced serious water contamination and toxic heavy metals pollutions.

There are twenty-nine known coal ash sites in Michigan, USA. Nineteen are within five miles of a Great Lake or a bay of a Great Lake, with several sitting directly adjacent to a bay. Three sites are within twenty miles of the Great Lakes. Every single coal ash impoundment or landfill sits on or near one of the state's lakes, rivers, streams, and wetlands. To improve the coal ash impoundments safety, a coal ash impoundments safety monitoring system (CAISM) was proposed based on the wireless sensor network (WSN) with the ability to monitor the impoundments stability, water level and the saturation levels remotely. This system has a robust Ad hoc network, which could adapt different kind of coal ash impoundments conditions.

2. CAISM Framework

2.1. The System framework

Some coal ash impoundments are rated for the degree of danger they pose to the communities and environments downstream. Dramatic failures aren't the only source of surface spills; smaller spills occur when impoundment dikes and dams leak less significant amounts, or impoundments overflow in heavy rains or floods.

On these conditions, the coal ash impoundments safety monitoring system (CAISM) aims to monitor the coal ash impoundments safety on anytime, in anywhere, which is based on the wireless sensor network (WSN). The entire system datum are acquired by all kinds of wireless sensors, processed by the cloud-computing platform, and then broadcasted to the terminals either on PDA or the other clients. The framework of CAISM is displayed in figure 2.

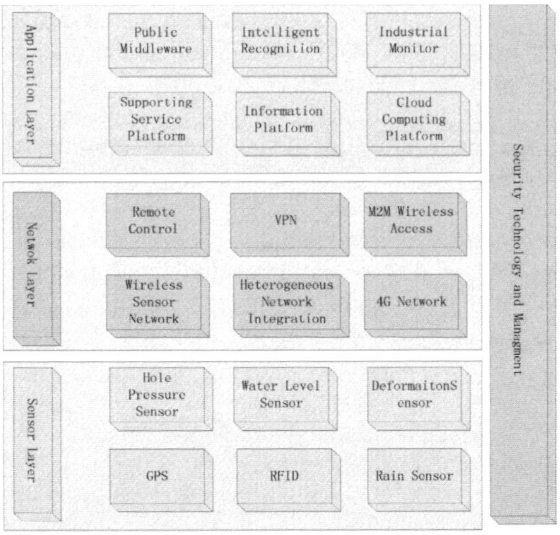

Figure 2. System framework of CAISM

2.1.1. Sensor Layer. Far more common than a coal ash dam break is the leaching of contaminants from ponds and landfills: the process by which toxic materials in coal ash dissolve in water and percolate through the earth. Leaching can expose people to dangerous toxicants at levels above safe drinking water standards.

Sensor layer of CAISM includes the hole pressure sensor, water level sensor, rain fall statistic sensor and the wind information sensor to monitor the key factors of the coal ash impoundments. Hole pressure sensors focus on the coal ash pond depth and dam saturated degree. Water level sensors monitor the water in the coal ash ponds, which could combined with rain fall statistic sensors. Especially, it is pivotal to calculate the water level according to the rainfall under heavy rainfall conditions. Coal ash also follows land and air pathways to result in human exposure. Coal ash disposal operations can generate dangerous quantities of airborne ash, due to mismanagement of both ponds and landfills. Coal ash is dangerous if inhaled, making fugitive dust a serious health concern. The health threat arises from minute particles of dust known as particulate matter, which may be composed of various substances. Wind sensors provide the information of the toxic coal ash direction, which are useful to judge the contamination area combined with the GIS map.

2.1.2. Network Layer. Coal ash is the second-largest industrial waste stream in the U.S., after mining wastes. [8] Coal ash is disposed in approximately 2,000 dump sites across the nation: at least 629 wet ash ponds [9] and 311 dry landfills at power stations, at least 100 offsite dry landfills, [10] and 750 inactive dumps, [11] and hundreds of abandoned and active mines. [12]

Network Layer of CAISM is a key component to transmit remote monitoring data back to control centre by using the wireless sensor network (WSN), 4G Ad Hoc network and M2M wireless access. Network layer makes sure the real-time data are encrypted under the security technology and managements. As the most of coal ash sites are relatively far from the city or community. WSN formed by a large number of sensor nodes where each node is equipped with a sensor to detect coal ash ponds safety features. The WSNs are regarded as a revolutionary information gathering method to build the information and communication system which will greatly improve the reliability and efficiency of infrastructure systems. Compared with the wired solution, WSNs feature easier deployment and better flexibility of devices.

IOP Conf. Series: Earth and Environmental Science **51** (2017) 012010 doi:10.1088/1755-1315/51/1/012010

2.1.3. Application Layer. The aim of CAISM is to provide key safety features to the users. The application lay of CAISM contains public middleware, intelligent recognition, information platform and supporting service platform.

The public middleware of CAISM is much easier to install on the user's devices. User friendly interface is important to show the monitoring results directly. Information platform broadcast up-to-date information on the health and safety related to the local residents lived near the coal ash impoundments. Supporting service platform provide a sponsored packet switched data service including receiving a request in a wireless network for a sponsored packet switched data service from a user, determining a sponsor for the requested service in accordance with stored policies, and monitoring a session between the user and the sponsor. The application layer of CAISM provide necessary services to ensure that effective communication with another application program in a network is possible.

3. Application of CAISM

The CAISM is a 24/7 automatic coal ash safety monitoring system. Considering the system safety, all the users are required the system security access validation. After checking the user's roles correctly, the main interface of CAISM is presented in figure 3. This is a coal ash impoundments with twelve saturated lines monitoring spots and one dam water level monitoring spot. The green labels that presented on the screen are the accumulated 3D deformation in millimeter unit and the pink labels that displayed on the screen are the saturated lines. Judging by the pre-alarm parameters, the alarm information will flash on the screen if any sensor value is beyond the safety pre-alarm value. The options of different selections help the users to choose the concerned monitoring category to show on the screen.

Figure 3. The main monitoring interface of CAISM

The CAISM real time partial water level monitoring data is shown in figure 4. The water level of the coal ash impoundments rises almost two meters in the past one month. In the late April, the water level increased almost one meter in about twelve days.

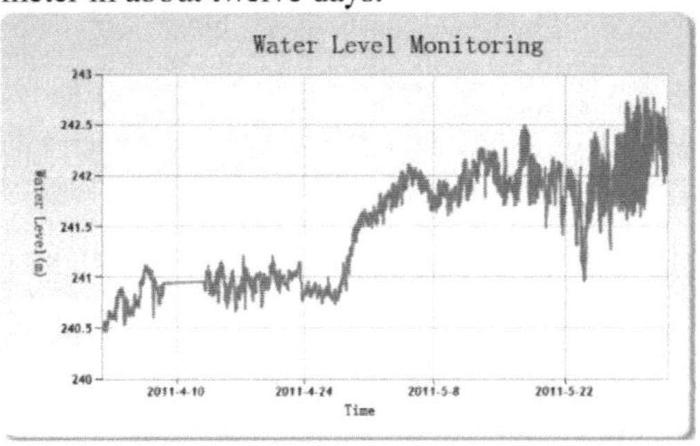

Figure 4. CAISM real time water level monitoring data

The storm and rain have great effects on the coal ash impoundment safety. In the rain season, flood regulation model is crucial to keep coal ash ponds safety. CAISM provides the flood regulation model as shown in following:

$$Q_{mp} = 0.278F\left[\frac{S_p}{\tau^n} - \mu\right] \quad (1)$$

$$\tau = \frac{0.278L}{mj^{0.333}Q_{mp}^{0.25}} \quad (2)$$

Where τ means the convergence time (h). Q_{mp} stands for the peak flow (m3/s). F is the drainage area (km2). 0.278 is the unit conversion coefficient. s_p is the rain power in twenty four hours(mm/h). L is maximum the flow length in the drainage. n is the rain decreasing coefficient. μ is the loss coefficient. m is the drainage coefficient. j is the average longitudinal slope.

According to the hydrographical manual, flood process shape coefficient $\beta = 3600\tau Q_p / W_p$, and the flood peak flow $W_p = 1000hF$. The five-hundred year period flood peak flow process Q_m.

As the flood regulation capacity is full-size, the flood water level increases are not obvious. The drain time is time-consuming because of the drain flow is a little low. For the five-hundred year flood regulation, as the water level raises 0.96m, the left beach length is 99m. After 24 hours drainage, the beach length will be 114.4m. After 72 hours later, the beach length will be 190.2m. The flood regulation result is displayed on figure 5.

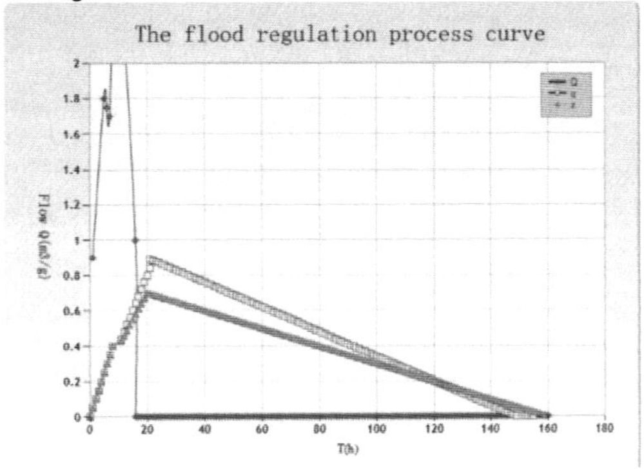

Figure 5. Flood regulation process cure, Five hundred years

Coal ash impoundments safety has a close relationship with the rainfall and drainage ability. The floods and landslides of coal ash impoundments that were triggered by the rainfall will cause a number of deaths and resulted in widespread damage to houses, roads and other infrastructure. CAISM can adjust the drainage according to the rainfall condition, which could avoid the failures and water pollutions.

4. Conclusions

Coal ash impoundments are widely spread all over the world as long as the coal-fired power plants are the vital energy source. The failures of ponds and leaching of contaminants have been proven to be of significant concern, which can produce serious ground water pollution. The toxic substances commonly found in coal ash, including mercury, lead, vanadium, selenium, and arsenic, are known to pollute water and could pose serious public health risk. Many disposal sites lack adequate safety regulation such as protective liners or groundwater monitoring and many of the impoundments are nothing more than earthen dams. It is highly risky to ignore the problem of structural instability of coal ash impoundments due to all the risk factors mentioned above. It is thus necessary to properly monitor the impoundments using a system similar to the one proposed in this paper.

Currently, there is no regulation for monitoring coal ash impoundments. The CAISM aims to monitor the coal ash impoundments safety based on wireless sensor network (WSN) on demand and remotely. It has the ability to monitor the key features of the coal ash impoundments. Combining with the rainfall sensors, CAISM provide the water level varieties under rainfall condition to prevent any kind of spill or failures. Coal ash impoundments safety is crucial to environment and surrounding ground water. CAISM needs to be further developed to integrate the toxic substances plume prediction, which will be more efficient to monitor the ground water pollution.

Acknowledgment
This work was supported by National Natural Science Foundation of China (71373245, 50974109), the basic research funding of China Academy of Safety Science and Technology (2016JBKY07), and National Key Technology R&D Program of the Ministry of Science and Technology (2016YFC0801305), and the 12th Five-Year National Science and Technology Support Program (2015BAK40B01, 2015BAK40B02)

References

[1] Sun E J, Zhang X K, Li Z X, Pei Y B. Simulation analysis on tailings dam saturated and semi-saturation infiltration of under rainfall conditions, *J of Safety Sci and Tech*, 2012, **8** (3):5-8.
[2] Feng H N ,Yang Y H ,Gong X N. Test research on engineering characteristic of fly ash. Rock and Soil Mechanics, 2002, **23** (5):579-82.
[3] Zhang B P, Dang J Q. Soil Mechanics and Foundation. *Xi'an:Xi'an Cartographic Press*, 2001.
[4] Liang X P, Su C D. The comprehensive utilization of powdered coal ash and the focal point of further development. *J of Hebei Institute of Tech*, 2005, **27** (3): 148-50.
[5] Tu G C. Test research and application of fly ash consolidation build dam. *Electricity Power Survey and Design*, 2003, (3):71-6.
[6] Sun E J, Zhang X K, Cheng S. Centrifuge and shaking Table Experiment on The Tailings Dam Failure. *China Safety Science Journal*,2012, **22** (6): 130-7.
[7] Sun E J, Nieto A, Li Z X, Kecojevic V.. An Integrated Information Technology Assisted Driving System to Improve Mine Trucks-Related Safety. Safety science, 2010, **48** (10):1490-97.
[8] Testimony of Lisa Evans, Attorney, Earthjustice, before the Subcommittee on Energy and Mineral Resources, Committee on Natural Resources, U.S. House of Representatives. June 10, 2008. http://www.earthjustice.org/library/legal_docs/ evans-testimony-emrsubcom.pdf.
[9] U.S. Environmental Protection Agency. Information Request Responses from Electric Utilities: Responses from Electric Utilities to EPA Information Request Letter: Database of Survey Responses. http://www.epa.gov/osw/nonhaz/industrial/ special/fossil/surveys/index.htm#surveyresults.
[10] U.S. Environmental Protection Agency. Regulatory Impact Analysis for EPA's Proposed Regulation of Coal Combustion Residues (CCR) Generated by the Electric Utility Industry. April 30, 2010 at 34.
[11] U.S. Department of Energy. Coal Combustion Waste Management Study, ICF Resources, Incorporated, February 1993 at page 1 of Executive Summary.
[12] Earthjustice. Waste Deep: Filling Mines is Profit for Industry, But Poison for People, February 2009, http://earthjustice. openissue.com/sites/default/files/library/reports/earthjustice_ waste_deep.pdf.

Green Innovation Design of Products under the Perspective of Sustainable Development

Guo yi-fei

Institute of Tianjin University of Science and Art, Tianjin, China

E-mail:519941822@qq.com

Abstract. with the continuous development and progress of productivity, product design spans with the age now and forms a diversified formation. But when there is a progress of science and technology, the speed of consumption of resources is also increasing, the relationship between mankind and the nature further worsens, resources will be exhausted, pollution is increasingly serious and shocking public nuisance events occur frequently. People pay more attention to the environment due to the harms to the nature caused by the development of industry, science and technology. Nowadays, people have clearly recognized the important role of design in environmental protection. The ecological research on constructing a new aesthetic relationship between human and nature has drawn wide attention from all circles of the society. Green innovation design has become the focus of global concern.

1. The role of green innovation design in sustainable development

Design is pushing the progress of mankind and caring the relationship between man and nature and creating a reasonable way of life. According to history, the choice of green design is inevitable, only green design can achieve the harmony of natural environment, social culture, and economic development, therefore, the development of green design is essential. First of all, green design is a basic guarantee of product quality. In terms of product design, green design is not only the principle of "3R" (reduce, recycle and reuse), but also to meet the green and healthy standards in many aspects of the product and prevent the environmental destruction and resource waste, such as product materials, functions, technology, quality, in addition to these aspects, the green design also pays more attention to the physical and psychological impacts of products on the users[1].

Secondly, in the economic aspects, green design also has its high value. People realize that environmental damage will affect their own living standards, the public awareness of environmental protection has been improved, while the Government enacts the relevant legislation, people are slowly looking forward to the green innovation design. With the people' in-depth understanding on green innovation design, the products designed by this method are accepted and chased after by people.

2. Principle of green design

Compared with the traditional design, the green design should follow the following principles:

2.1 Principle of making full use of resources

The principle of making full use of resources means giving priority to consider the properties and renewability of the resources when selecting and using resources, which is consistent with the concept of sustainable development and the waste cannot be caused by improper design. Second, in the designed products, try to use the products in the entire life cycle farthest. As shown in Figure 1.

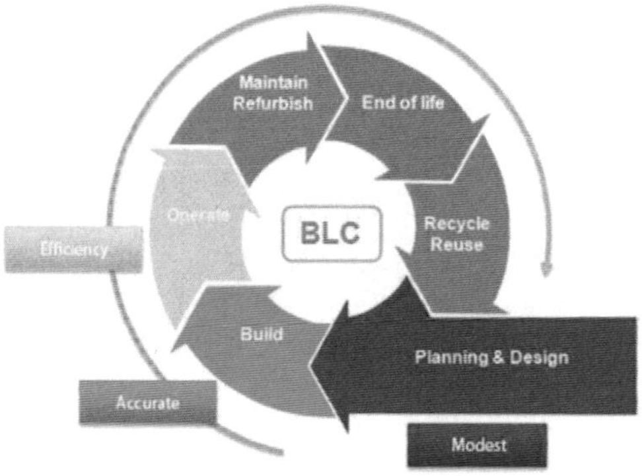

Figure 1.The whole life cycle of products

2.2 Principle of energy conservation
The principle of energy conservation refers to selecting renewable energy sources, such as wind energy and solar energy, rather than fossil fuels, such as coal and oil, making the designed products consume the fewest energy sources in the whole life-cycle.

2.3 "Zero pollution" principle
When designing any products, consider how to eliminate pollution, avoid the end treatment method of treatment after pollution, develop new products by the strategy of prevention first, governance second, and control pollution from the root.

2.4 Principle of advanced technology
Green design must make the product be in "green" color, requires using the most advanced technology, and requires designers to be creative, so that the product has the best market competitiveness.

2.5 Principle of the best ecological and economic benefit
Green design should not only consider the economic benefits created by the product, but also start from the point of view of the sustainable development to consider the impacts of the environmental behavior of products in the life cycle on ecological environment and society as well as the loss of environmental and ecological benefits and social benefits. In other words, the green design should make the green product producers not only obtain the good environmental benefit, but also obtain the good economic efficiency, namely the best ecology economic efficiency.

3. The inevitability and necessity of green design
The irreversible era development direction of the 21st century is the green design, which is a brand new product design philosophy. The emergence of green design is not accidental, it is inevitable and necessary. In the process of product development, it is the only way and it is the inevitable choice to ensure human survival and achieve sustainable development; and it is the inevitable product under green consumption wave in the new century and the foothold of the future enterprises [2]. In order to properly implement green design, we must explore new approaches to industrial design and follow certain guidelines, by which the environmental protection can be integrated into industrial design better. In addition, it must be recognized that green design is still in the formation stage of the development process. In order to improve it, more designers are needed to know it, understand it to solve the various problems in the systematic process. As shown in Figure 2.and 3.

Figure 2. Bamboo material products

Figure 3. Bamboo material products

In the past industrial design, there is a lack of green design concept, when designing products, the designers do not consider the conservation of materials and recycling of materials and the result is that the non-renewable resources on earth are gradually swallowed. With the rise of postmodernism, product modeling pursues individuation excessively, the product shapes are exceedingly strange, so that the product manufacturing is more difficult and the utilization of materials reduces, and a lot of resources are wasted. In addition, driven by commercial design, product replacement cycle is shorter and shorter, thus accelerating the depletion of non-renewable resources on earth. Therefore, in order to alleviate the resource crisis and energy crisis on earth and take full advantage of the limited resources on earth, and industrial design can walk on green design road only [3].

First of all, green design is the inevitable choice of self-improvement of industrial design concept. In its formation and development process, the industrial design constantly improves, and improves its own shortcomings. But no matter how industrial design concept evolved in the past, it did not consider the environmental factors, namely, in the design, the green concept is not integrated, which is its greatest shortcoming and indirectly brings some negative impacts on human beings. We know that for any kind of product, no matter how long its life is, but it cannot avoid the dumped ending caused by ageing and damage. Because there is no green consciousness, the previous design results that the product is abandoned and non-recyclable, non-reusable, the product wastes have accumulation. Also because there is no green design sense, some toxic materials or the materials difficult to be degraded will be selected in the design, so that the product wastes will pollute the environment. With the explosion of world's population, the speed of generating household refuse and product waste is faster and faster. It is reported that since the 1990s, about 1 billion tons of household wastes has been generated in the world, which is almost 3 times of the weight of the entire human. And in the treatment process, there may be secondary pollution. In the new century, in order to fundamentally solve the excessive garbage, reduce pollution, the way is to take the green design road in industrial design only.

Secondly, the green innovation design is necessary to guarantee the sustainable development. In modern times, the science and technology develop rapidly, the productivity improves in a fast speed, and the unprecedented material and culture treasures are created, thus the human civilization reaches a new stage. To conform to this irreversible time trend, for the industrial design, a subject need to be perfected constantly, the idea of sustainable development must be blended in the product design process in order to combine the ecology, environment and economic development as an organic whole with reason as goal, in the economic development, it is necessary to consider the long-term bearing capacity of natural ecological environment, make sure that the environment and resources can not only meet the economic development requirements, but also be regarded as one of the factors for the survival of mankind to directly meet the long-term survival demands of human, in order to form a comprehensive development strategy that the resources and energies can be effectively used, and environmental pollution can be reduced to minimum extent, and all of these can be realized by developing on the green design road in the industrial design. So green design is necessary for the sustainable development of human society, and is a vital part of sustainable development. The green design is an inevitable product under the trend of green consumption in 21st century. In the 1970s, with the deterioration of environment, the people's environmental awareness improved constantly, especially in 1980s, the "Green Movement" with the topic of environment protection was carried out in the globe gradually. In western countries, the "green food"-based green revolutions were broken out one by one. Under the effect of green food, the green products such as green refrigerator, green computer, green auto and green buildings come out in succession. The green consumption aimed to improve the quality of life was born at the right moment. The green consumption is a pattern of consumption to follow health, uphold nature and be beneficial to environment protection. When people are on shopping, they not only concern the function, life, type and price of the products, they even pay more attention to the environmental performance, and prefer to buy the green products with more costs[4]. As shown in Figure 4 and 5.

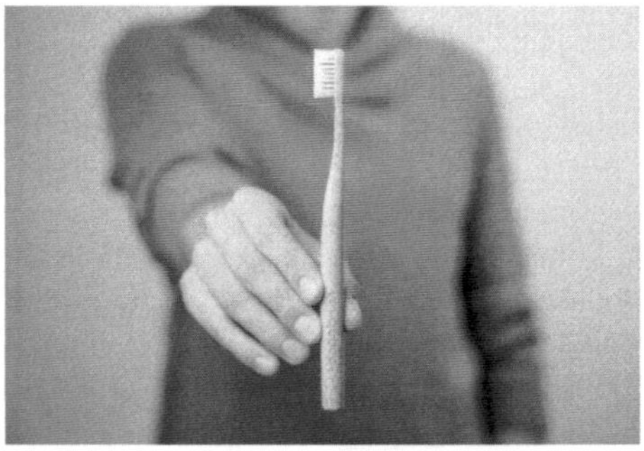

Figure 4. Biodegradable toothbrush

Figure 5. Biodegradable boxes

In addition, people consume the resources and energies in the life with the purpose not only to live in comfort, but also to save resources and energies in quantity. An oversea consumption survey shows that there are more than 70% of American express that the green image of enterprises and the products may directly affect the purchasing behavior: in European market, nearly half of the customers will consider the environmental issues while buying the products, and 94% of Italians favor the green products, and 82% of the Germans will pay attention to the environmental factors while consuming the products, in Asia, such as Japan, Korea and Hong Kong, the customers are also wild about the green products. It is thus clear that the green consumption has become the trend in the international market. To adapt this trend, the designers must blend the environment protection in the product design, in order to meet the environment protection requirements in material and design, meanwhile, combine the green property of packing material, finally obtain the green products, which is the meaning of green design.

Finally, the green design is the basis of the enterprises in the future, and the effective approach to eliminate the green trade barrier. With the rapid development of information technology, the world seems to be smaller, and the globe changes to be the global village in broad sense, and the concepts of domestic market and foreign market also become indefinable. During the process of economic globalization, the enterprise must get into international market if it wants to develop in the future. The precondition of getting into international market is that the products must meet the international standards and specifications. With the arising of green consumption in the globe, various countries in the world have formulated demanding green product standards, if some enterprises produce the products which are not meeting the standards, as a result, they will be removed from the world market relentlessly, and finally, the enterprises will go to die[5].

4. Green innovation design promotes the sustainable development

With the development of human society, especially the accelerated industrialization process and expanded urban size, the regional pollution, ecological damage and resource and energy exhaustion become serious and endanger the human's survival and sustainable development. Considering the demands of information society in the 21st century and the consumer demands under the condition of population expansion and resource reduction, in the future, the design shall be aimed at the healthy, hygienical, safe and easily operational equipment, thus the parts of equipment or products can be easily replaced and reused, in order to save resources and energies as much as possible, and reduce the environment pollution. The green design conforms to the history development tendency, focuses on using the resources effectively, improving the product quality, and reducing the waste amount caused as industrial and agricultural production activities and various vital movements, which is a design method meeting the environmental requirements. In a word, the green design can save energy and resource, can relieve the environment pollution, and realize the healthy and harmonious development of society, economy and environment, and there are perfect environmental benefits.

5. Conclusion

As a new design method, the green innovation design is widely accepted in western developed countries, after it fully displays effect in enterprises, the objective ecological and economic benefit can be obtained. Asia, Japan and Korea have conducted many researches, China also conducts some researches on green design and green manufacturing, and meanwhile, the Natural Science Foundation of China and 863 High-technology Research Project also subsidizes some scientific research projects, including green design and green manufacturing, and resource optimization, which have been applied in some green product design. As a new design method, it is not mature yet in theory, and there are still some works to be done; so it is necessary for the enterprise circles and academic circles to pull together in order that it can be widely used in the product design.

Reference

[1] Hu ai-wu,Fu zhi-hong. 2003 Green Design for Products. *ournal of Zhuzhou Institute of Technology* 17(5): 21-23.
[2] Ai fei. 2012 *Based on the principles of green design product*. Wuhan:Wuhan University of Technology.
[3] Xu xi-hua. 1999 Ecological design: Theinevitable chioce of the development of the industrial design.*Journal of Zhejiang university*, 29(4): 132-136.
[4] An li-li, Yang ming-lang, Wu xiao-hua. 2007 Analysis and Conception of Innovation Design of Products. *Packaging Engineering*.28(7): 185-187.
[5] Yang chao-nan, Sun wei. 2015 New Materials and Innovative Design Method Combined with Sustainable Design. *Light Industry Science and Technology*.4:114-116

IOP Conf. Series: Earth and Environmental Science **51** (2017) 012012 doi:10.1088/1755-1315/51/1/012012

A Study on China's System of Public Participation in Environmental Protection

----Taking Haikou's "Two Constructions" for Example

Wang Xi[1]

[1] Business Administration Department, Hainan College of Economics and Business, Haikou, Hainan, China

Abstract: In recent years, China's environmental condition becomes more and more serious and all kinds of pollution incidents happen frequently. Although government departments have implemented various policies and measures to curb the deterioration of the environment, public participation in environmental protection is the key solution of which the breadth and depth determines the level of environmental protection. This paper first analyzes the problems of public participation in environmental protection, such as defects of current legislation, low level of environmental awareness and participation, weak environmental NGOs etc. and then puts forward three ways to optimize the system of public participation in environmental protection by the example of Haikou City's "Two Constructions" Project, that is, perfect the relevant legal mechanism, increase publicity and support environmental NGOs. Finally the paper makes a summary and outlook of China's system of public participation in environmental protection.

1. Current situation and problems of public participation in environmental protection activities in China

Public participation in environmental protection means allowing, encouraging and supporting the public to actively participate in environmental decision-making and management, evaluate and supervise the relevant activities of government in order to realize public interests. Public participation in environmental protection activities in which the government plays a leading role helps to make up for the limitations of government work which may lead to a failure, to stimulate the enthusiasm of the public, and to form a good atmosphere of all people contributing themselves to environmental protection. However, there still exist many problems in the course of practice.

1.1 Defects in laws and regulations of pubic participation in environmental protection

The laws and regulations of China's environmental protection are still immature compared with that in developed countries. It has a certain relationship with China's national conditions, economic development level and people's livelihood. Moreover, some of the environmental laws and regulations in our country are issued in answer to the serious or emergent events, which is bound to affect the effectiveness and implementation of them. The public has a right to know, to supervise and to participate in the environmental protection activities, but the definition of specific procedures, rules and methods in the existing laws and regulations is relatively vague, because of which the public cannot exercise their rights and participate in the practice. Besides, people still don't know how to act in accordance with the laws in order to get an effective result.

1.2 Low level of awareness and participation in environmental protection

Environmental pollution has come home to roost in recent years which can be easily felt by the human beings. The haze, air pollution and bad weather phenomenon gradually infiltrate into the lives of ordinary people. It is a pity that people haven't put themselves into environmental protection activities while they keep complaining about the disgusting environmental condition. Part of the public is not aware of the importance of contributing to environmental protection. And what's worse, they even can not consciously abide by environmental laws and regulations. All of this results from poor propaganda of public participation in environmental protection to a certain extent.

1.3 Weak environmental NGO (ENGO)

NGO has a characteristic of folk and independence compared with government departments. As the third party and strength besides the government and the market, it is an important bridge and channel for government to strengthen its ties with people through effective communication. At present there are many non-governmental organizations active in the field of environmental protection, such as Friends of Nature, Global Village of Beijing, Green Homes, China Small Animal Protection Association and so on in China, which help to promote the popularity of environmental protection activities to a certain extent. However, because of China's vast territory, the overall strength of ENGO is relatively weak. The number, funding, capabilities and the impact of ENGO are very limited.

2. Measures of optimizing the system of public participation in environmental protection

In our country, corresponding measures, policy and mechanism is indispensable as to make sure that the system of public participation in environmental protection really works.

2.1 Improve the laws and regulations of public participation in environmental protection

All things must be done by law. In developed countries, public participation makes up one fundamental principle in environmental laws[1]. In China, there also exist relevant laws and regulations.

For example, the fifth part of Law of Environmental Protection of the People's Republic of China clearly states the public's right to know, to participate and to supervise; further consolidates the status of public participation in assessing the environmental impact of construction projects as well as establishes the Environmental Public Interest Litigation System. It is not hard to see that our country pays a due attention to public participation in environmental protection. But that is not enough. In the future, it is rather important to define the boundaries of civil rights and obligations in environmental protection activities and point out the ways, forms and specific procedures of public participation to make sure that there are fewer legal and technical obstacles for people to help change current bad environmental condition[2].

Haikou's "Constructing A National Civilized City and National Sanitary City" (Two Constructions) is a quick response to Premier Li Keqiang's call for "Mass Entrepreneurship and Innovation " at the

World Economic Forum held in Sept. 2014. Then the implementation of *Haikou Patriotic Health Regulation* was passed by a unanimous vote at the 38th Session of the Standing Committee of the Fifteenth Haikou Municipal People's Congress on Apr.27th, 2016. To overcome the heavy difficulties in urban sanitary management, the regulation makes special stipulations to the management of the questionable community, standards for the dining industry, tobacco and smoke control, the safety of secondary water supply and so on. It also makes a clear division of responsibilities, highlights the system design, implements strict conduct, strengthens the legal responsibility, and lays emphasis on the actual effect of law enforcement, providing legal insurance for the municipal project. It is worth mentioning that the law puts up a lot of new perspectives which support "Two Constructions". For instance, in the third paragraph of article 33, it says that for residential area without department in charge, property manager or owner committee, municipal and district people's government could provide encouragement and support by means of substitute subsidies with rewards and hardship subsidies, so that residential communities could set up owner committee or owner's self-rule organization, employing the department in charge or full-time environmental workers to achieve public sanitary management. Moreover, article 26 explicitly stipulates who are in charge of the sanitary management in the area: construction supervising units or construction units; the right user or supervisor of the idle state-owned land; the owner, the right user or supervisor of the idle collective land[3]. These regulations guarantee the public their legal status in environmental activities, offer approaches for participation, as well as arouse the enthusiasm of the public.

2.2 Enhance public awareness of environmental protection

The government, as the leader of environmental protection activities, must strengthen the publicity of environmental protection activities, cultivate public awareness of environmental protection and form an atmosphere of "Everyone works for environmental protection". Then how can we enhance public awareness of protection? The point is simplifying the terminology of environmental protection activities and providing practical opportunities for public participation[4]. First, for ordinary people, they find it is rather difficult to understand those obscure technical terms written in environmental protection materials and thus gradually lose their interest in relevant activities. Therefore, simple and familiar vocabularies should be used so that the attractiveness and influence of environmental protection events can be fully enhanced. What's more, blog, audio-visual electronic magazine, micro-blog, WeChat, mobile phone client and other new media has witnessed rapid development, which can bring about remarkable result if properly used. Second, as one old saying goes, real knowledge comes from practices, more practical opportunities should be provided for the public. In this way, all people, from child to the aged, can get to know how to participate in various environmental activities.

Haikou's "Two Constructions" creates many ways to strengthen public awareness of environmental protection. Citizens can ask questions, make comments and give advice with the help of modern media network. For instance, one TV program named "Liangjian" offers people an opportunity to make their voice heard even dissect government work. Besides, in another action--I know, I participate, I contribute-- people from all walks of life put their heads to find more effective solutions to environmental pollution[5].This is a platform for people to make environmental contributions and welcomed by the whole society. Through a variety of rich and colorful propaganda activities, more and more community residents begin to join the volunteer team, which proves that public awareness of participation in environmental protection has been enhanced.

2.3 Strengthen the power of ENGO

First, financial support is essential for ENGO to work in environmental protection activities. A lot of folk organizations turn out to be formalistic because of lack of funds. All-China Environment Federation released a "Development Report of ENGO" in 2006 which recorded more than 70 percent ENGOs were lack of a fixed source of funding, 60 percent had no special offices, and 40 percent full-time personnel had no salaries[6]. Therefore, the government's funding is the most important support for ENGO. Government departments can provide funds for legitimate ENGOs through financial allocations or they can appropriate policies to encourage public, corporate and celebrity donations to environmental public welfare undertakings. In addition, the government should do a series of business support for ENGO like legal consultance, technical training, etc., to create a favorable atmosphere for them to take part in environmental protection. Second, the government should vigorously promote public participation in the field of environmental protection, because volunteers are the most important human resources in ENGO. Financial support and human resources get together to lay a solid foundation for the development of ENGO. Besides, China's ENGO should actively learn from their foreign counterparts advanced experience, establish internal supervision mechanism, increase transparency to win public trust and attract more people to join in environmental activities.

There are a few but not many environmental non-government organizations (ENGO) in Hainan, among which Nature Conservation & Development Research Association of Hainan Province is correspondingly famous. Nevertheless, the "Two Constructions" activity will surely provide theoretical basis and practical experience for the establishment and development of ENGO in Haikou. There are numerous activities while here only representative examples are presented for brief analysis. For instance, to improve environmental awareness and promote ecological civilization, Haikou municipal government conducted a selection aiming to find "The Most Beautiful Sanitation Figures" on the Hainan Environment Day, Jun.5th, 2016. The contestants should make outstanding contributions to ecological civilization construction and environmental protection, actively participate in the "E&I" project as well as the city construction[7]. The activity is undoubtedly a strong impetus to the public, the volunteers and non-governmental organizations. It not only shows the government's strong support, but also improves their motivation to participate in environmental activities. Furthermore, in the "Two Constructions" activity, civilian efforts emerge in large numbers. HNA organizes 1500 volunteers who spare no effort to salvage dead fish in the East Lake， wipe out the small ads on the Telegraph post and clear away rubbish in urban areas. Elderly volunteers go to community residents committees and set up the health supervision team. They increase health inspections, sweep the streets, correct uncivilized behaviors, and many community residents are influenced by their enthusiasm and kindness[8]. Following the lead of non-governmental organizations and volunteers, Haikou citizens devote to environmental protection activities, and thus public participation has been further improved.

3. Summary and outlook

Haikou's "Two constructions" has achieved good results in environmental protection, which may set up a good example for other cities and towns to change the environment deterioration. However, one most important solution to environment problem lies in public participation. As mentioned above, China's public participation system is not perfect and needs to go through a long process of exploration. But as long as the government and the public make concerted efforts to constantly improve the corresponding legal system, enhance public awareness of environmental protection, strengthen the power of ENGO and explore other effective ways, the environmental pollution will be controlled and sustainable development will be achieved.

References

[1] Zhuo Guangjun 2012 Public Participation in Environmental Protection in China Ph D Thesis. Chongqing: Chongqing University 42-48
[2] Cheng Zijun, Li Zhiqiang 2009 On the Problems and Reasons of Public Participation in Environmental Protection in China Environmental Science and Management 43 -46
[3] Liu Gong 2016 Haikou's "Two Constructions" Is Still on the Road... Hainan Daily 2
[4] Li Jing 2015 Study on Public Participation in Environmental Protection in China Ms D Thesis. Xi'an: Northwest University 30-35
[5] Wang Mingchu 2016 Make People's Life Better. Hainan Today 2
[6] An Investigation on the Present Situation of China's Environmental NGO 2006 www.sina.com.cn
[7] Li Qiang, Wang Youwen, Fan Xiaodong 2016 "Two Constructions" Is Changing Our City. People. Rule of Law 2 15
[8] Ji Yanling 2016 Haikou's Folk Organizations Is Working Hard in "Two Constructions". Southern Metropolis Daily News 9 22

"Decoupling" Indicators for Evaluation of Urban Low-Carbon Economy Development: a Case Study of Shanghai

Jing Zhao[1,2], Ting Xu[3], Daqiang Yin[3], Weihua GU[1,2] and Qing Huang[1,2]

[1]WEEE Research Centre of Shanghai Polytechnic University, 201209 Shanghai, China
[2]Shanghai Collaborative Innovation Centre for WEEE Recycling, 201209 Shanghai, China
[3]Key Laboratory of Yangtze River Water Environment, Ministry of Education, College of Environmental Science and Technology, Tongji University, 201209 Shanghai, China

E-mail: zhaojing@sspu.edu.cn

Abstract. Environment are closely related to energy and they are two of the most important issues this century. So the development of a low-carbon economy and the construction of environment friendly society are the only way for the further development of China its own. In this study, "decoupling" indicators for evaluation of a low-carbon economy development are proposed. "Decoupling" indicators of energy consumption and economic growth were selected to reflect the response relationship between economic growth and the changes of resources, including *weak*, *strong* and *expansive/recessive* degree of decoupling. Similarly, "decoupling" indicators of carbon dioxide emissions and economic growth were chosen to reflect the response relationship between economic growth and the changes of environment. Shanghai, in the Eastern China, was selected as a special case. The results showed a change from *expansive negative coupling* to *week decoupling* regarding both of energy consumption and carbon emissions in Shanghai during the period 1991-2013, indicating that we have achieved some significant outcome in the process of the low-carbon economy development in Shanghai. However, in view of the possibility of *expansive negative coupling*, we should continue to make efforts for the development of the low-carbon economy.

1. Introduction

Environment are closely related to energy and they are two of the most important issues this century. It is reported that China was responsible for nearly three-quarters of growth in global carbon emissions from the burning of fossil fuels and cement production between 2010 and 2012 [1, 2]. However, researchers re-evaluated China's carbon emissions and the results showed that China's CO_2 emissions from fossil fuel combustion and cement production is 14 per cent lower than the emissions reported by other prominent inventories in 2013, which is 2.49 gigatonnes of carbon [1, 3-5]. Anyway, as a responsible country China have an obligation to make strategies for carbon emission reduction as a response to climate change. If China cannot make efforts to reduce carbon emissions, the future economic and social development will be hampered. Given this, the development of a low-carbon economy and the construction of environment friendly society are the only way for the further development of China its own.

2. A theoretical framework on decoupling to evaluate the development of the low-carbon economy

Although China has developed low-carbon economy in recent years, there were no criteria and indicators to measure it. The "decoupling" indicators of the development of a low-carbon economy was a good try. Referred to the conceptual model of decoupling and coupling [6, 7], "decoupling" indicators of energy consumption and economic growth was selected to reflect the response relationship between economic growth and the changes of resources in this paper, including three indicators: Gross Domestic Product (GDP), the total amount of energy consumption (TEC) and the energy consumption per unit of GDP (TEC/GDP). Similarly, "decoupling" indicators of carbon dioxide emissions and economic growth were chosen to reflect the response relationship between economic growth and the changes of environment, including three indicators: Gross Domestic Product (GDP), the total amount of carbon dioxide emission (CO_2) and the carbon dioxide emission per unit of GDP (CO_2/GDP).

The category of "decoupling" indicators of carbon dioxide emissions and economic growth is the same as "decoupling" indicators of energy consumption and economic growth (Table 1). In here, take "decoupling" indicators of energy consumption and economic growth as example. Decoupling (TEC/GDP<0)) can be divided to three subcategories: in *weak decoupling*, GDP and TEC both increase, *strong decoupling* occurs when GDP grows and TEC decreases and *recessive decoupling* when GDP and TEC both decrease. Similarly, negative decoupling (TEC/GDP>0) includes three subcategories: in *expansive negative decoupling* GDP and TEC both increase, in *strong negative decoupling* GDP decreases and TEC increases and *weak negative decoupling* occurs when both variables are decreasing.

Table 1. The degrees of coupling and decoupling of energy consumption (△TEC) from economic growth (△GDP) and the degrees of coupling and decoupling of carbon dioxide emissions (△CO2) from economic growth (△GDP)

Degree	△TEC	△CO_2	△GDP	△(TEC/GDP)	△(CO_2/GDP)
Strong decoupling	<0	<0	>0	<0	<0
Week decoupling	>0	>0	>0	<0	<0
Recessive decoupling	<0	<0	<0	<0	<0
Strong negative decoupling	>0	>0	<0	>0	>0
weak negative decoupling	<0	<0	<0	>0	>0
Expansive negative decoupling	>0	>0	>0	>0	>0

3. Statistical data on decoupling in Shanghai

Both of GDP and TEC data were from Shanghai statistical yearbook (2014) [8]. In order to eliminate the effect of currency fluctuations GDP in Shanghai was recalculated based on the base price in 1978. Then the comparable price was obtained and used to calculate the growth rate of GDP. The unit of the total energy consumption was ten thousand tons of standard coal, which did not involve the price factor. So these data can be used directly to calculate the growth rate of the total energy consumption.

3.1. The response relationship between economic growth and energy consumption in Shanghai

The results showed a change from *expansive negative decoupling* to *weak decoupling* regarding energy consumption during the period 1991-2013 in Shanghai (Table 2, Figure 1). In view of the growth rate of GDP and energy consumption, energy consumption elasticity coefficient was changed between 1991-2013. The energy consumption elasticity coefficient was 1.22 in 1991, indicating that the growth rate of energy consumption growth was faster than that of economic growth. Then in the following years, the value of energy consumption elasticity coefficient was lower than 1 (for example, the lowest value was 0.11 in 2012, and 0.39 in 2013).

Table 2 The indicators of energy consumption, carbon dioxide emissions and economic growth in Shanghai 1991-2013

Year	ΔTEC	ΔCO_2	ΔGDP	Δ(TEC/GDP)	$\Delta(CO_2/GDP)$
1991	275.46	215.51	46.10	0.07	0.05
1992	190.40	148.96	102.30	-0.41	-0.32
1993	289.81	226.73	120.04	-0.29	-0.22
1994	229.91	179.87	132.59	-0.33	-0.26
1995	289.23	226.28	149.77	-0.26	-0.20
1996	160.34	125.44	156.87	-0.31	-0.25
1997	132.61	103.75	173.23	-0.30	-0.24
1998	115.29	90.20	157.14	-0.22	-0.17
1999	245.08	191.74	175.14	-0.14	-0.11
2000	380.29	297.52	204.61	-0.09	-0.07
2001	395.30	309.26	216.61	-0.08	-0.06
2002	354.56	277.39	257.81	-0.12	-0.10
2003	547.00	427.94	312.09	-0.08	-0.06
2004	609.30	476.68	404.85	-0.11	-0.09
2005	819.41	641.06	371.02	-0.01	-0.01
2006	650.65	509.03	460.50	-0.10	-0.08
2007	794.75	621.77	621.19	-0.12	-0.09
2008	536.91	420.05	456.68	-0.08	-0.06
2009	160.02	125.19	423.40	-0.12	-0.09
2010	833.75	652.28	575.63	-0.04	-0.03
2011	69.35	54.26	505.52	-0.13	-0.10
2012	91.67	71.72	500.06	-0.11	-0.08
2013	341.52	267.19	551.89	-0.07	-0.05

Data resource: 《Shanghai statistical yearbook 2014》 [8]

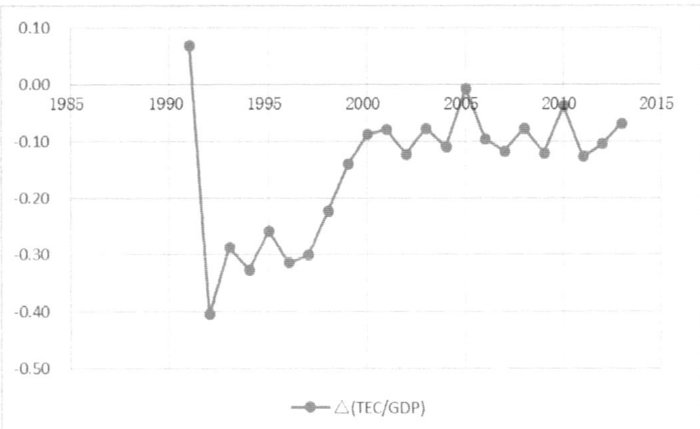

Figure 1 Decoupling of energy consumption from economic growth in Shanghai 1991-2013

Overall, that the low energy consumption growth rate can support rapid economic growth may be due to the following two reasons: one is increased energy efficiency benefited from the technological progress and regime innovation; the other is optimized and upgrading industrial structure. Excluding the impact of price factor, according to the comparable price in 1990, the total energy consumption per ten thousand yuan GDP continued to reduce, and the energy efficiency in Shanghai was significantly higher than the national average value.

3.2. The response relationship between economic growth and environment stress in Shanghai

3.2.1 Calculation of carbon emissions. Because there were no direct monitoring data of carbon emissions in China, most of the researches were based on the energy consumption to obtain the carbon emissions data. Based on the existing research results of carbon emission estimation at home and abroad [9-11], the carbon emissions in China are estimated by the following formula:

$$\mathrm{Ems} = \alpha \times \mathrm{E} \qquad (1)$$

Where Ems is the amount of carbon dioxide emissions (unit: tons of carbon equivalent), α for carbon dioxide emissions factors and E for the amount of energy consumption (unit: tons of standard coal). According to the calculation method of CO_2 emissions from fossil fuels combustion by Oak Ridge National Laboratory (ORNL), α is 0.7194.

Based on the hypothesis that there was a positive correlation between carbon emissions and energy consumption, and considering the difference in the energy structure in different areas and the carbon emission factors in different energy types, the carbon emissions in different provinces was estimated by the following equation:

$$\mathrm{Emsi} = \frac{\mathrm{Ei}}{E} \times \frac{\gamma i}{\gamma} \times \mathrm{Ems} \qquad (2)$$

Where Emsi and Ei are the amount of carbon dioxide emissions and the amount of energy consumption in *i* province respectively; Ems and E are the total carbon dioxide emissions and the total energy consumption in China respectively; γi for the average of carbon dioxide emissions factors in *i* province and γ for the average of carbon dioxide emissions factors in China. In this paper, γ is 0.8, γ_{sh} is 0.87[12].

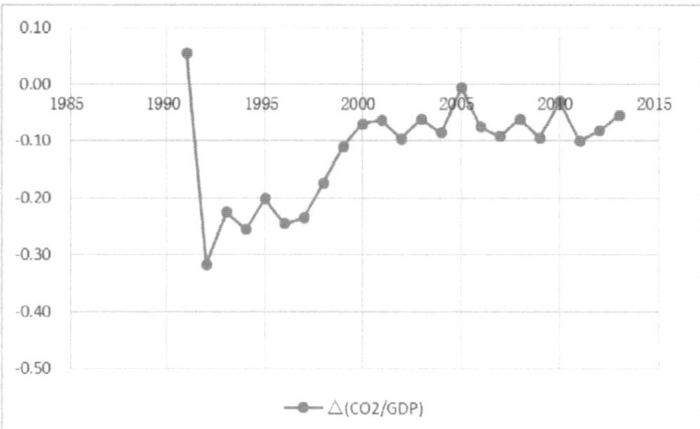

Figure 2 Decoupling of carbon emissions from economic growth in Shanghai 1991-2013

3.2.2 The response relationship between economic growth and energy consumption in Shanghai. The data showed a change from *expansive negative decoupling* to *weak decoupling* regarding CO_2 emissions between 1991 and 2013 in Shanghai (Table 2, Figure 2), which was consistent with the results of decoupling of energy consumption from economic growth. With the rapid development of the economy in China, △(CO_2/GDP) was greater than 0, indicating that the annual growth rate of CO_2

was greater than that of GDP in 1991. Fortunately, Shanghai experienced *weak decoupling* of CO_2 emissions from GDP between 1992 and 2013. However, as the growth rate of CO_2 emissions in Shanghai increased much more quickly than that of GDP in recent years, there will probably be change from *weak decoupling* to *expansive negative coupling*.

4. Conclusions

In summary, the statistics showed a change from *expansive negative coupling* to *week decoupling* regarding both energy consumption and carbon emissions in Shanghai between 1991 and 2013, indicating that we have achieved some significant outcome in the process of the development of the low-carbon economy in Shanghai. The growth rate of energy consumption and carbon emissions were both lower than that of economic growth, and economic growth did not exhibit a positive correlation with energy consumption or environment stress during the period 1992-2013, suggested that *week decoupling* from GDP to energy consumption and carbon emissions. However, in view of the possibility of *expansive negative coupling*, we should continue to make efforts for the development of the low-carbon economy.

Acknowledgements

The authors gratefully acknowledge the financial supports from the National Natural Science Foundation of China (No.21507080 and 41501598), Project for Training of Young Teachers in Colleges and Universities of Shanghai City (No. ZZZZEGD15018) , the Key Subject of Shanghai Polytechnic University (WEEE recycling and pollution control, XXKZD1602). The authors also thank the Shanghai Collaborative Innovation Centre for WEEE Recycling for funding and support.

References

[1] Boden T A, Marland G, Andres R J. Global. 2013 *Regional, and national fossil fuel CO2 emission*. Oak Ridge National Laboratory, US Department of Energy.

[2] Liu Z, Guan D B, Crawford-Brown D et al. 2013 *A low-carbon road map for China Nature* 500 143-145.

[3] Olivier J G, Janssens-Maenhout G, Peters J A. 2013 *Trends in global CO2 emission: 2013 report PBL Netherlands Environmental Assessment Agency*.

[4] National development and reform commission. *Second national communication on climate change of the People's Republic of China*. Department of climate change (in Chinese).

[5] Liu Z, Guan D B, Wei W, et al. 2015 *Reduced carbon emission estimates from fossil fuel combustion and cement production in China Nature* 524 335-338.

[6] Tapio P. 2005 *Towards a Theory of Decoupling: Degrees of Decoupling in the EU and the Case of Road traffic in Finland Between 1970 and 2001 Transport Policy* 12 137-151.

[7] Vehmas J, Kaivo-oja J, Luukkanen J. 2003 Comparative de-linking and re-linking analyses of material flows in the EU-15 member countries. *ConAccount Conference "Quo Vadis MFA?"*. Wuppertal, Germany.

[8] Shanghai Municipal Statistics Bureau. 2014 *Shanghai statistical yearbook*. Beijing: China statistics press (in Chinese).

[9] Oak Ridge National Laborary. Carbon Dioxide Information Analysis Center. *Estimates of CO2 emission from fossil fuel burning and cement manufacturing*. Oak Ridge: ORNL, 1990.

[10] Guo R, Cao X J, Li Y K et al. 2009 Shanghai's strategies for carbon emission reduction as a response to climate change. *Journal of Tongji University (Natural Science)* 37 515-519 (in Chinese).

[11] Qian J. 2004 *The study of Metropolis carbon source and carbon pool: a case study in Shanghai*. Shanghai: East China normal University (in Chinese).

[12] Wang Z, Zhu Y B. 2008 Study on the Status of Carbon Emission in Provincial Scale of China and Countermeasures for Reducing its Emission. *Bulletin of Chinese Academy of Science*, 23 109-115 (in Chinese).

The Influence of Older Age Groups to Sustainable Product Design Research of Urban Public Facilities

Zhang Wen-juan and Song Hou-peng

Tianjin University of Technology, No. 391 Binshui Road, Xiqing District, Tianjin City

E-mail: songhoupeng1990@gmail.com

Abstract. Through summarize the status quo of public facilities design to older age groups in China and a variety of factors what influence on them, the essay, from different perspective, is designed to put forward basic principle to sustainable design of public facilities for the aged in the city, and thus further promote and popularize the necessity of sustainable design applications in the future design of public facilities for elderly people.

1. Introduction

Under the progress of social development and the promulgation of social security as well as various policies at home, the concern pertinent older age group has been raising increasingly. In such a trend, public systematic services to this group have also become a focus people pay more attention to, especially the importance of equalization and rationalization of urban public facilities distribution to those old people who live in the city. With the promotion of government public service function and its social management ability, public facilities design and its spatial planning in first-tier cities have been exerting guided and exemplary effect to ones in second and third-tier cities. Compared with developed countries, however, there is significant difference in the construction of public facilities for older age groups and its systematic perfection at home, of which the applied effectiveness of public infrastructure and product iteration problem have became some obstacles needed to be solved that currently influence on urban sustainable development. By systematic research and analysis, it is difficult not to find out these root causes that design concept of public facilities product fails to keep pace with the times, the function of facilities product cannot satisfy old people's physiological and psychological requirements, and neglect the relationship of public facilities in between human, environment and nature. Therefore, to research and think the problems is a reality which must be resolved and faced to urban sustainable development in the future.

2. Product design and sustainable design of public facilities

2.1. Formatting the title Public facilities product design

Public facilities product is various type of service-oriented and public one offering to citizens, the function of which is designed to meet users' requirement, make life easier and serve the public. With rapid development of urbanization at home, public facilities product plays an increasingly important part in city image and window to foreign countries, so it is incumbent responsibility and obligation, for every designer, that reasonably grasp the design of public facilities product for urban sustainable development in the future. Generally speaking, good public service product design must cater to different user group's demand (particularly for some special social groups, such as elder people, children and disabled people, etc), and take spatial environment, natural landscape and other regional features as the basis of the background, at which it can not only concordantly coexist with people when product integrate into natural ecology environment, but also deliver design concept full of historical cultural atmosphere and humanistic concern to every target audience when urban charm has been improved constantly.

2.2. Public facilities sustainable design

In the development phase of modern urban planning and design, sustainable design, extended from the concept of sustainable development, has become a something new in public's awareness. On the whole, people's correct choice for the modern way of life is driven by the idea of sustainable development, which could mean that it not simply is the product of human's reflection to environmental and ecological damage, but is the recurrence of ethnic and responsibility in the entire society--- of which old people have been a key group concerned mostly with by it, as people can see, in today's society, public facilities product has already been an essential part of daily life in elderly people [1]. In general, public service product designed with ergonomics and fitting for older age groups can make the elderly people's life more convenient and secure, and sustainable public facilities product made by the combination of regional characteristic and local scenery can maximally reduce environmental pollution caused by products, which is not only helpful to product iterative refinement but to the reduction of resource and energy consumption [2]. In the current times, new design pattern of sustainability is built on the basis of human-environment-nature common for sustainable development. With the elevation of awareness in public to natural resource and ecological environment, it is better able to enhance old men's living quality in the city, make them to follow the development trend of social and environmental sustainability, and improve their living environment in the cities when sustainable public facilities product in accordance with aged group's needs to product safety and portability. Therefore, it is the only way in further promoting urban planning and the explorative direction in transforming sustainable development of new cities in China future that settle down to research the design system of public facilities product with sustainable concept.

3. Sustainable design factors of public facilities product for older age groups

Public facilities product involves with the very broad content ranges and influential factors, on account of it is a type of public one facing and serving for social mass. Given that, in the construction of theoretical system of public facilities product based on the design concept of sustainability, both variable factors in design processes and old users' physiological and psychological requirements have to be taken into consideration for the sake of its perfection. This section, from the beginning of regional characters, will make further discussion--- by the combination of local natural environment and humanism elements, and the integration of peculiarities in sustainable design including safety, portability, identifiability and cost-effective (Figure 1)--- that how to ameliorate the influence of public facilities in mass.

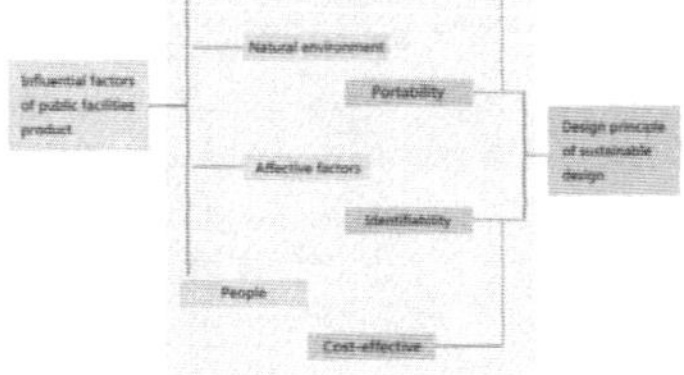

Figure 1. Diagram between public facilities and sustainable design

3.1. Regional factors in the design of public facilities product
The influence in regional culture is mainly from people's customs in daily life and their inherent cognitive thinking, but diverse characteristic in regional culture is primarily from different geographical positions. As for essence, regional characteristics is the epitome of regional culture, the reason of which its influence was formed in the long run of historical years and the accumulation of every generation, its profound cultural connotation and historic value exert inestimable effect in this day as well. From the viewpoint of aesthetic theory, it is prevailing trend to public facilities product design of modern city that integrate regional characteristic into sustainable design conception of product. For urban development, on the one hand, this type of product combined traditional elements with modern principle can not merely show S&T strength advancing with the times in modern urban public service, but present historical implication in urban for public consciousness through specific-image product design; for target audiences in cities, on the other hand, it is not difficult incorporating this comprehensive public product into their daily life as its design philosophy comes from regionally cultural character; for aged people, in particularly, public facilities product characterized by local culture tends to make the resonance with their psychological perception. As a whole, the design pattern with recognizability is better for the identifying operation of products in older age groups when it closed the distance between products and old users, of course, which is also the reason why it is easier for old men in the use of them. Simultaneously, this type of public product, when it showed the city charm for foreign countries, could convey urban historical flavor, humane concerns and the concept of humanization design, and thus further accelerate sustainable development of the city. "Columbus Circle", for example, that is the dominant landmark and tourist attraction in Manhattan of New York City. As shown in the figure 2, on the basis of local people's adoration for sport and light, designer made the construction of a circular building by the design concept of concentric circle, which not only made it become central area of the city, but evolve into "center of the universe" in people's inner feelings.

Figure 2. Columbus Circle in Manhattan

3.2. Human factors in the design of public facilities product
Human has always been actor and undertaker in social activities, and their requirement has always been the basic element in system of them, because of this objective reality, audience's demand has become the entry point of practical problem that need to be solved primarily by designers in the process of design for public facilities product. In line of steady development of today's society, aging level in China has increasingly been aggravating. As is statistic, gross population at home has grown beyond 1.36 billion, of which there are about 210 million ones over 60 years old, and who are 15.44 percent of its total population. By all accounts nothing could be more convincing than figures, thus, older age group could once again be back on the table of holistic urban development with the growth of the authentic data, which means that to regard them as the main focus of public facilities product design is an important presentation to promote public service system more complete [3].

Older age group, as one of the important measure of social progress, has been playing significant role in urban development. Most often, they have their own specific demand and idea to public facilities product under the drive of their life and practical experience, so, for purpose of working out a public product corresponding to city image and receivers' demand as well, it is extremely important to collect and accept their opinion and thought. In a general way, requirement in older age group can be divided into physiological one and psychological one. As they grow older, old men's bodily various system functions have gradually been declining, and this senile limitation also makes them generate some psychological barriers in use of cumbersome and complicate products. Based on the actual situation, it is necessary thing during product design that, according to local characteristics, designers should make the investigation of physical condition in older age group and the solicitation for proposal in them to public service products, which could transform old people's role in public service system from product user to participant in preliminary design of the product [4]. From the point of view, to make human as one of standard in social public, on one hand, could better improve practical efficiency, reduce resource wastes and fulfill the ideas of sustainable development; on the other hand, for older age group, this concept could give full play of positive effect on the recognition of self ability and the construction of mental quality in them, simultaneously, their own social value in urban development would also be more and more prominent with its guide.

3.3. Natural environmental factors in the design of public facilities product

The relationship model between human and environment has long been the topic which is being researched and discussed in academic community. As part of nature, human is closely associated with natural environment in term of subsistence as they originated and evolved from here [5]. Since ancient times, human progress has been achieved at the price of consuming natural resources, hence, whatever economic development or social advancement in today, they have been ascribed to the scope of resources over consumption and ecological destruction. As can be seen, the imbalance relationship between human and environment cannot possibly be swept under the carpet for good. On the contrary, its drawbacks must have stood out with time providing, and human have also become the ultimate responsibility undertakers in consequence caused by unpredictably natural disaster. Therefore, to rationally plan the public service system and guide the connection between human, environment and ecological pace in city are the key breakthrough for improving the relationship model between human and environment.

Figure 3. Terra grass arm chair

As the media of the harmony between man and environment, public facilities product has to be designed in the consideration of natural environmental factors. From the point of view of essence, natural environment is the foundation of urban construction, and mutualism between both sides is the goal of urban comprehensive development. Given that, in order to create a harmony in public facilities product and natural environment, it is imperative that, in the light of natural environment characteristics at different place, combine design material with nature and draw on local resources at the prerequisite of maintaining the existing landscape. Furthermore, the public facilities product endowed with urban natural characters is not only helpful to accelerate the regenerative cycle of city ecosystem and strengthen citizens' environmental awareness, but to lead receivers to build up right view of nature and ecological value in their use of product to some extent. In a word, only keeping men and environment in public system in equal and interactive development pattern, will entire ecosystem develop stably and sustainability. "Terra grass arm chair", for example, designed by Nucleo is an efficient product that can perfectly blend in natural environment. As shown in the figure 3, the arm chair was constructed with cardboards, filled with soil and decorated with vegetation seeds, moreover, its own evolution, in the drive of specific design principle, would be going on in comply with natural rhythm.

3.4. Emotional factors in the design of public facilities product

"Emotion", as an emotion-dominated inner consciousness in human, can transmit perceptive effect to the surrounding people and environment in the form of behavioral language and body language, in the same way, corresponding things can exert certain influence upon human emotion as well. In a sense, it is very necessary in the design of sustainable product that research the changing process of mental emotion in elderly people, because their vulnerable inner state and stereotypical mind-set have very strong side effect on their cognition to products with age. However, in view of the existing emotional demand in old men, shape, structure, color and other visual factors should also be applied into the design of public product for catering for their need in product safety consciousness, at which older age group's initiative would be stimulated in use of public product, and their emotional needs would be extended with new design principle as well.

4. Concluding remarks

With the development of cities and the improvement of people's living standard, there has been profound changing in people's living habits, ways of thinking and behavioral consciousness. When people pursue the product function on materials, its spiritual one cannot also be ignored categorically, for which, in satisfying the people's (especially older age group) needs of work, leisure and recreation, the communication of emotional spirit and the inheritance of historical connotation are given too much importance in sustainable design of public facilities. So, only integrating into local feature in city, does public product designed with sustainable concept develop continuously in their local area. All in all, to make the construction of sustainable design meeting modern demand, in ways of discovering region cultural connotation and natural ecological characteristics in urban, must have made public facilities product with sustainability to really promote the sustainable development of urbanization.

References

[1] David Bergman. Sustainable Design A Critical Guide [M]. Xu Xin-lian, Chen Ran, Translate. Nanjing: Phoenix Science Press, 2013 p 32.
[2] Nathan Shedroff. Design is the Problem: The Future of Design Must be Sustainable [M]. Liu Xin, Tan Jing-yan, Translate. Beijing: Tsinghua University Press, 2011 p 89.
[3] Ye Qing. The Development of the Modern Urban Planning and Urban Design [D]. Shanghai: Tongji University, 2007 p 18.
[4] Yu Fei. Discussion on User-friendly Policy in the Design of Public Facilities [J]. Packaging Engineering, 2011(12): p 134-138.
[5] Carlo Vezzoli, Ezio Manzini. Design For Environmental Sustainability [M]. Liu Xin, Yang Hong-jun, Tan Jing-yuan, Translate. Beijing: National Defense Industry Press, 2010 p 67.

Identification of Bacterial Species in Kuwaiti Waters Through DNA Sequencing

K Chen

American International School of Kuwait, Hawalli, Kuwait

Ninachen575@gmail.com

Abstract: With an objective of identifying the bacterial diversity associated with ecosystem of various Kuwaiti Seas, bacteria were cultured and isolated from 3 water samples. Due to the difficulties for cultured and isolated fecal coliforms on the selective agar plates, bacterial isolates from marine agar plates were selected for molecular identification. 16S rRNA genes were successfully amplified from the genome of the selected isolates using Universal Eubacterial 16S rRNA primers. The resulted amplification products were subjected to automated DNA sequencing. Partial 16S rDNA sequences obtained were compared directly with sequences in the NCBI database using BLAST as well as with the sequences available with Ribosomal Database Project (RDP).

1. Introduction

Bacteria are key microbes in marine microbiology that play fundamental roles in the ocean environment. Apart form being the base of the food chain, supplying more than half of the world's oxygen, controlling the role of marine nutrients, and being major processors of greenhouses gases, they are also rapidly responsive to changes in their ecosystem, making bacteria an ideal indictor of marine change [1][2]. One such case is the microbialization of bacteria in contaminated waters over the past decade. The growing pathogenicity of marine microbes can cause disease in both marine and human life. Upon contact or consumption of specific bacteria, humans can experience cramps, nausea, diarrhea, tuberculosis, legionella pneumonia, typhoid, and cholera [3]. Bacterial diseases of fish include-but are not limited to: hemorrhagic septicemia, vibriosis, and columnaris [4]. The susceptibility of infection from pathogenic bacteria reinforces the importance of water purification before industrial, agricultural, or domestic applications. It is also ideal that water treatment processes use purification methods that limit leftover waste, as brine sediments often contain pathogens not killed off during treatment, and disposing them back into the ocean reintroduces the harmful bacteria into the aquatic environment in abundance.

There is growing interest in marine bacteria in the scientific world because pathogenic, or otherwise, their and identification serves many purposes. Firstly, extremophiles can be used as a measure of water pollution levels or aquatic condition because these species only grow in extreme environments. For example, halophilic bacteria only thrive in waters of high salt concentration, and are an indicator of overly high salinity. Next, being aware of pathogenic bacteria and locations were specific genera reside allows researchers, scientists and doctors to better prevent disease that could be contracted in the area-some of which can be very dangerous. Treatment plants can also better allocate purification methods for more effect water treatment with an understanding of the bacterial genera residing in specific marine regions. For example, bacteria killing methods of purification-such as electrochemical remediation, will be more efficient in treating water samples containing pathogens than other methods that do not eliminate bacteria. Employing target specific ways of water purification is more effective, efficient, and economical. Finally, bacterial identification is also important because many bacterial species such as: Marinobacter hydrocarbonoclasticus, Carboxydothermus hydrogenoformans, and Sphingomonas harbor properties that can be industrially useful in pharmaceutical, biotechnological, agricultural, chemical, and environmental fields.

The purpose of this study is to investigate the diversity and properties of marine bacteria species from Salwa Beach (N1), North West (N2) and Sulaibikhat Beach (N3) by directly culturing seawater samples, PCR amplification of 16S rRNA gene, and DNA sequencing.

2. Methodology

2.1. DNA Growth and Preparation

Bacteria from each beach was first grown by spreading 3μL of each sample onto separate marine agar plates, then incubated at 28°C in an bacteriological incubator for 24 hours. Grown bacteria cultures were then collected and mixed into cell pellets of 10μL of sterile distilled water. 48 samples of bacteria were collected from each beach.

Cultivation was also attempted on MacConkey, and nutritional agar plates. However, results indicated that bacteria could not be cultured on said plates, which made the identification of fecal coliforms based on culturing conditions and microscopical morphologies impossible.

2.2. PCR

Bacterial 16S rDNA was amplified from the extracted genomic DNA by using the following universal eubacterial 16S rRNA primers: forward primer 5' AGAGTTTGATCCTGGCTCAG 3' and reverse primer 5' ACGGCTACCT TGTTACGACTT 3'. An ABI PCR System 9700 was used to perform the PCR protocol in a volume of 24 μL of bacteria strain identification solution mix containing 10 x 2.5μL buffer, 1.25μL magnesium chloride, 1μL dNTP, 0.05μL primer 1, 0.05μL primer 2, 0.1μL enzyme taq, and 19.05μL of distilled water per 1μL of template bacteria. Thermal cycling consisted of a 10 minute initial denaturation at 95°C, 30 cycles of denaturation at 94°C for 30 seconds, 30 seconds of annealing at 60°C, and extension for 2 minutes at 72°C with a 10 minute final maintenance at 4°C [5].

2.3. Agarose Gel Electrophoresis

From each PCR product, an aliquot of 15μL was subject to 1.0% agarose gel electrophoresis containing 0.5 x TBE buffer (pH 8.0), and 5μL of ethidium bromide was used to stain the gel. Electrophoresis was conducted at 100 Volt, 400mA for 20-30 minutes with 0.8μL of 100bp DNA marker. After, the DNA bands were observed under ultraviolet light using gel documentation system [5].

2.4. Sequencing of 16S rRNA Genes

Sequencing reaction was carried out using ABI PRISM 310 Genetic Analyzer (PE applied Biosystems). For the sequencing reaction, Big Dye Ready Reaction DyeDeoxy Terminator Cycle Sequencing kit (Perkin-Elmer) was employed.

2.5. Data Analysis

Sequence analysis was performed with sequences in the NCBI database using BLAST as well as with sequences available with Ribosomal Database Project (RDP)

3. Results and Discussion

3.1. Salwa Beach

Total bacteria counts: 85

IOP Conf. Series: Earth and Environmental Science **51** (2017) 012015 doi:10.1088/1755-1315/51/1/012015

Table.1. Identified bacteria isolates in Salwa Beach after data analysis.

Isolate	Identified Bacterial Strain
N1-001	Uncultured bacterium clone ncd981d10c1
N1-002	Exiguobacterium acetylicum
N1-003	Cronobacter sakazakii
N1-004	Uncultured bacterium isolate KWSUB08_76 16S
N1-005	Marinobacter sp. MMRF630 16S
N1-006	Alteromonas macleodii
N1-007	Marinobacter litoralis
N1-009	Marinobacter sp. HOT2G5 16S
N1-010	Alteromonas sp. NBF17 16S
N1-011	Alteromonas macleodii
N1-012	Bacterium SCSIO13055 16S
N1-013	Bacterium SCSIO13055 16S
N1-014	Pseudoalteromonas sp. Bac192 16S
N1-015	Alteromonas macleodii
N1-016	Alteromonas sp. SS12.5 16S
N1-017	Phaeobacter caeruleus
N1-018	Enterobacter cloacae
N1-019	Alteromonas macleodii strain CSB14KR 16S
N1-020	Sphingomonas
N1-021	Spirometra erinaceieuropaei
N1-022	Alteromonas sp. OCN004 16S
N1-023	Hydatigera taeniaeformis
N1-024	Gamma proteobacterium
N1-025	Schistosoma rodhaini
N1-026	Erythrobacter citreus
N1-027	Exophiala mesophila 3-isopropylmalate dehydrogenase
N1-029	Tropicibacter sp. MCCC1A07686 16S
N1-030	Xanthomonas oryzae pv. Oryzae
N1-031	Pseudoalteromonas sp. F497 16S
N1-032	Pseudomonas sp. X3-2
N1-033	Pseudomonas psychrophila
N1-034	Colacium sp. Ungok033107C
N1-035	Citromicrobium bathyomarinum
N1-036	Candida catenulate
N1-038	Diphyllobothrium latum
N1-041	Schistosoma rodhaini
N1-045	Spirometra erinaceieuropaei
N1-046	Tetrapisispora phaffii
N1-047	Plasmodium falciparum

Table.2. Pathogenic bacteria found in Salwa Beach isolates after data analysis.

Isolate	Description
N1-003	Cronobacter sakazakii: causes cronobacter disease with fatality rates of 50-80%. Infants and immunocompromised adults are most susceptible [6].
N1-018	Enterobacter cloacae: Enterobacter cloacae infection has the highest mortality rate amongst all Enterobacter infections. It causes bacteremia, skin and soft-tissue infections, urinary tract infections, endocarditis, intra-abdominal infections, septic arthritis, osteomyelitis, CNS infections, ophthalmic infections, and lowers respiratory tract

	infections [7].
N1-021	Spirometra erinaceieuropaei: is a tapeworm that causes a parasitic infection called sparganosis [8].
N1-023	Hydatigera taeniaeformis: a parasitic tapeworm with cats as their primary definitive hosts [9].
N1-025	Schistosoma rodhaini: commonly known as blood-flukes, they are parasitic flatworms that cause schistosomiasis in humans-an infection considered by the World Health Organization to be the second most socioeconomically devastating parasitic disease infecting hundreds of millions worldwide [10].
N1-030	Xanthomonas oryzae pv. Oryzae: causes bacterial blight, a serious disease for rice, grass, and sedges worldwide [11].
N1-038	Diphyllobothrium latum: is the largest human tapeworm and the principle species of Diphyllobothrium causing diphyllobothriosis It infects humans through the somsuptiono of raw or undercooked seafood, but can also affect fish and other mammals [12].
N1-041	Schistosoma rodhaini: refer to N1-025
N1-045	Spirometra erinaceieuropaei: is a tapeworm that infects humans and domestic animals. In humans, is causes sparganosis [8].
N1-047	Plasmodium falciparum: is a protozoan parasite that is the deadliest of five human malaria species, being responsible for more than 85 % of all malaria cases, and has the highest mortality rate amongst all malaria species [13].

It can be inferred that Salwa beach is home to an abundance of tapeworms, as 6 out of 8 pathogenic bacteria identified were of such. Tapeworms, which generally reside in areas of inadequate sanitation is an indication of the area's contamination levels. Salwa beach is also a harbor. Many tapeworms could dwell inside the intestines of fish, so extra persuasions should be taken when eating undercooked seafood to prevent infections.

Table.3. Bacteria with industrial uses found in Salwa Beach isolates after data analysis.

Isolate	Description
N1-020	Sphingomonas: typically found in environments polluted with toxic compounds, they have the ability to utilize the contaminants as nutrients. Their biodegradative and biosynthetic capabilities have been utilized for numerous biotechnological applications including bioremediation of environmental contaminants for the production of extracellular polymers such as sphingans [7]. It is also effective in degrading degrade alkylated polyaromatic hydrocarbon (2-methylphenanthrene), commonly known as polystyrene plastic [14].
N1-031	Pseudoalteromonas sp. F497: capable of synthesizing biologically active molecules that produce compounds beneficial to eukaryotic organisms [15].
N1-046	Tetrapisispora phaffii: killer yeast that secretes a glycoprotein known as Kpkt that has antimicrobial activity with spoilage yeasts under winemaking conditions. It can be used to combat contaminating wild yeasts in food, to control pathogenic fungi in plants, and to develop novel antimycotics for the treatment of human and animal fungal infections in the medical field [16].

There is no pattern to the useful bacterial species identified in Salwa beach, but the particularly useful Sphingomonas was cultured and identified. Its unique ability to synthesize and degrade toxic compounds in water could be applied as an effective organism in water purification-a function that can be utilized in Kuwait's many water treatment plants.

3.2. North West Beach

Total bacteria counts: 112

Table.4. Identified bacteria isolates in North West Beach after data analysis.

Isolate	Identified Bacterial Strain
N2-001	Muricauda sp. SA1 16S
N2-002	Alteromonas macleodii strain G7C_40m_M_02 16S
N2-003	Myxococcus fulvus
N2-004	Marinobacterium stanieri
N2-005	Idiomarina aquimaris
N2-006	Marinobacter hydrocarbonoclasticus
N2-007	Spongiibacter marinus
N2-010	Flavobacteriales bacterium
N2-011	Marinobacter sp. NNA5 16S
N2-012	Alteromonas macleodii
N2-013	Corynebacterium humireducens
N2-014	Rhodobacteraceae bacterium
N2-015	Bacillus pumilus
N2-016	Schistosoma mattheei
N2-017	Muricauda sp. mur1 16S
N2-018	Flavobacteriales bacterium
N2-019	Marinobacter hydrocarbonoclasticus
N2-020	Marinobacter flavimaris
N2-021	Rhodobacteraceae bacterium
N2-022	Marinobacter Pelagius
N2-023	Rhodobacteraceae bacterium
N2-024	Loxodonta africana ALG12, alpha-1,6-mannosyltransferase
N2-025	Pectobacterium sp. SCC3193
N2-026	Brassica napus
N2-027	Schistosoma curassoni
N2-028	Marinobacter flavimaris
N2-029	Erythrobacter citreus
N2-030	Marinobacter hydrocarbonoclasticus
N2-036	Novosphingobium sp. SG8916S
N2-037	Marinobacter sp. G5-4a
N2-038	Rhodobacteraceae bacterium
N2-039	Mus musculus
N2-040	Solitalea canadensis
N2-041	Pseudoalteromonas sp. ECSMB85 16S
N2-042	Marinobacter flavimaris
N2-043	Spongiibacter marinus
N2-044	Schistosoma curassoni
N2-045	Cyanobium sp. NIES-981
N2-046	Marinobacter sp. DY57-1 16S
N2-047	Uncultured organism clone ctg_CGOF252 16S
N2-048	Roseovarius pacificus

Table.5. Pathogenic bacteria found in North West Beach isolates after data analysis.

Isolate	Description
N2-016	Schistosoma mattheei: a parasite of sheep, goats, and cattle in regions of West Africa [17].
N2-027	Schistosoma mattheei: refer to N2-016
N2-044	Schistosoma mattheei: refer to N2-016

Schistosoma mattheei was the singular pathogenic bacteria culture and identified in North West Beach, and it has no direct threat to either humans or marine life. The area of the collected sample should be taken into consideration, as it is the closest to a direct polluting source (industrial power plant) with notably less fish than in popular fishing regions such a Salwa Beach. This could possibly explain the sample's lack of pathogenic bacteria, as many species of pathogens reside in marine life.

Table.6. Bacteria with industrial uses found in North West Beach isolates after data analysis.

Isolate	Description
N2-006	Marinobacter hydrocarbonoclasticus: are important to bioremediation because they can effectively degrade hydrocarbons-especially those found in alkanes and major components of oil as they are tolerant of high salinity and can grow aerobically and anaerobically [18]. The marinobacter hydrocarbonoclasticus' distinctive ability makes it ideal for treating recalcitrant pollutants [19].
N2-015	Bacillus pumilus: is used as an active ingredient in agricultural fungicides because its growth on plant roots prevents rhizoctonia and fusarium spores from germinating [20]. Bacillus pumilus proteases are also used in chemical, detergent, food, and leather industries as antimicrobials and antifungals.
N2-019	Marinobacter hydrocarbonoclasticus: refer to N2-006
N2-029	Erythrobacter citreus: is a potential candidate for removing poisons because it can resist and reduce the toxic compound tellurite [21].
N2-030	Marinobacter hydrocarbonoclasticuss: refer to N2-006

3 isolates of Marinobacter hydrocarbonoclasticus was found out of the 48 bacteria samples taken, suggesting that the species is abundant in North West Beach waters. Its properties are very useful in purifying Kuwaiti water, as it is able to degrade hydrocarbons found in oil. Oil is an abundant resource in Kuwait that frequently pollutes the county's water sources, but is difficult to remove once contaminated. Therefore, the potential of using bacteria to purify this recalcitrant pollutant is exciting and could perhaps be more cost efficient than currently employed methods.

3.3. Sulaibikhat Beach

Total bacteria counts: 126

Table.7. Identified bacteria isolates in Sulaibikhat Beach after data analysis.

Isolate	Identified Bacterial Strain
N3-001	Carboxydothermus hydrogenoformans
N3-002	Lactococcus garvieae
N3-003	Vibrio alginolyticus
N3-004	Vibrio parahaemolyticus
N3-005	Vibrio sp. H-188 16S
N3-006	Virgibacillus sp. SK33 16S
N3-007	Vibrio alginolyticus
N3-008	Vibrio alginolyticus
N3-010	Exiguobacterium undae
N3-012	Uncultured gamma proteobacterium
N3-013	Exiguobacterium undae
N3-014	Exiguobacterium undae
N3-015	Exiguobacterium undae
N3-017	Vibrio sp. 3677
N3-018	Vibrio alginolyticus
N3-019	Vibrio sp. B2-15-2 16S
N3-020	Vibrio sp. H-219 16S
N3-021	Vibrio sp. 3683
N3-022	Vibrio orientalis
N3-023	Vibrio natriegens

N3-024	Providencia rettgeri
N3-025	Clupea harengus translocase
N3-026	Anas platyrhynchos
N3-027	Drosophila takahashii
N3-028	Vibrio sp. sw9 16S
N3-029	Uncultured bacterium clone I3Q1XXJ01APUOI 16S
N3-030	Nippostrongylus brasiliensis
N3-031	Vibrio sp. 3677
N3-033	Vibrio alginolyticus
N3-034	Vibrio sp. B2-15-2 16S
N3-035	Vibrio sp. H-219 16S
N3-036	Vibrio sp. 3683
N3-038	Vibrio orientalis
N3-039	Vibrio natriegens
N3-040	Providencia rettgeri
N3-041	Clupea harengus
N3-042	Anas platyrhynchos
N3-043	Drosophila takahashii
N3-044	Vibrio sp. sw9 16S
N3-045	Uncultured bacterium clone I3Q1XXJ01APUOI 16S
N3-046	Nippostrongylus brasiliensis

Table.8. Pathogenic bacteria found in Sulaibikhat Beach isolates after data analysis.

Isolate	Description
N3-002	Lactococcus garvieae: is a zoonotic pathogen that causes infective endocarditis, spondylitis, osteomyelitis, and liver abscesses in humans. But mostly affects the Japanese yellowtail, rainbow trout, and grey mullet [22].
N3-003	Vibrio alginolyticus: like most Vibrio infections, this is a foodborne disease that causes vibriosis, otitis and wound infection [23].
N3-004	Vibrio parahaemolyticus: also cause vibriosis, along with gastrointestinal illness when ingested [24].
N3-007	Vibrio alginolyticus: refer to N3-003
N3-008	Vibrio alginolyticus: refer to N3-003
N3-018	Vibrio alginolyticus: refer to N3-003
N3-030	Nippostrongylus brasiliensis: is a gastrointestinal roundworm that primarily infects rats [25].
N3-033	Vibrio alginolyticus: refer to N3-003
N3-040	Providencia rettgeri: causes infections involving the urinary tract with symptoms of polyuria, hematuria, fevers, and dysuria. It has also recently been implicated as etiologic agents in traveler's diarrhea [26].
N3-046	Nippostrongylus brasiliensis: refer to N3-030

Species of bacteria from the Vibrio genus was especially prevalent in Sulaibikhat Beach, with 6 out of 10 identified isolates fitting into the category. Both Vibrio alginolyticus and Vibrio parahaemolyticus, along with Providencia rettgeri, are all foodborne infections. Therefore, precautions should be taken against eating undercooked food fished from this area.

Table.9. Bacteria with industrial uses found in Sulaibikhat Beach isolates after data analysis.

Isolate	Description
N3-001	Carboxydothermus hydrogenoformans: is an extremely thermophilic bacteria that feeds almost entirely on toxic carbon monoxide and water, and produces hydrogen gas as part of its metabolism [27]. Its ability to rapidly and efficiently conduct this conversion

	makes it highly valuable for the future where hydrogen gas is being studied as potential biofuel [28].
N3-013	Exiguobacterium undae: According senior research engineer Wei-Min Wu in his report "Biodegradation and Mineralization of Polystyrene by Plastic-Eating Mealworms. 2. Role of Gut Microorganisms", injections of exiguobacterium undae into the gut of mealworms allows them to survive solely on a diet of polystyrene, and thus, can significantly help in the biodegradation and mineralization of plastic [29].
N3-014	Exiguobacterium undae: refer to N3-013
N3-015	Exiguobacterium undae: refer to N3-013
N3-034	Vibrio sp. B2-15-2 16S: is capable of producing biosurfactant that can be applied into environmental, technical, and potentially pharmaceutical industries [30].

A wide diversity of environmentally useful bacterial species was identified in Sulaibikhat Beach. The most abundant was Exiguobacterium undae. Its property of reverting mealworms to feed solely on polystyrene has enormous potential as an eco-friendly way of degrading plastic. Next, Carboxydothermus hydrogenoformans's diet of only toxic carbon monoxide and water, which later produces hydrogen gas, could provide an alterative for oil. Lastly, a useful species of Vibrio, sp. B2-15-2, was identified amongst its pathogenic counterparts and demonstrates the impressive biodiversity of marine bacteria and their respective properties.

4. Conclusion

The data analysis revealed that Kuwaiti water samples harbor extremely diverse bacteria belonging to over 53 bacterial genera. 24 genera were isolated in Salwa Beach, 17 in North West Beach, and 12 in Sulaibikhat Beach. In Salwa Beach samples, identified genera consisted primarily of Alteromonas (17.5% out of 40), North West Beach samples consisted of prominently Marinobacter (26.8% out of 41 sequences), and Sulaibikhat Beach samples consisted of mostly Vibrio (46.3% out of 41 sequences). Alteromonas and Marinobacter strains were found to be abundant and distributed in both Salwa and North West Beach, while none were cultured and identified in Sulaibikhat Beach.

Each sampled location exhibited specific patterns of cultivated pathogen bacterial species that reveals information regarding each beach's unique marine condition and life. In Salwa Beach, most pathogens were tapeworms, only Schistosoma mattheei were isolated from North West Beach, and 3 pathogenic species of Vibrio was found in Sulaibikhat Beach. Since tapeworms and Vibrio species are ubiquitous in aquatic environments copious in marine life, the intense mariculture in Salwa and Sulaibikhat Beach may be partially accountable for the high occurrence of these pathogens.

On the other hand, identified industrially useful bacteria demonstrated no pattern in accordance to sampled location, but similarities in valuable property. While a wide range of diverse, unrelated bacterial cultures were isolated in each beach, the majority of all had biodegradative and biosynthetic capabilities. For example, Sphingomonas found in Salwa Beach can utilize toxic compounds and contaminants as nutrients, Marinobacter hydrocarbonoclasticus found in North West Beach can effectively degrade hydrocarbons, and Carboxydothermus hydrogenoformans found in Sulaibikhat Beach can feed almost entirely on toxic carbon monoxide and water, while producing hydrogen gas. Upon further study, these species-along with the others identified, potentially offer incredible scientific breakthroughs in the usage of microorganisms in pharmaceutical, biotechnological, agricultural, chemical, and environmental fields.

In summery, the purpose of this study was to cultivate, isolate, identity, and determine properties of marine bacterial species in Kuwait from Salwa Beach (N1), North West (N2) and Sulaibikhat Beach (N3) through PCR amplification of 16S rRNA gene, and DNA sequencing of bacteria grown from collected water samples. The findings indicate towards a high level of macrobiotic diversity, and shed light on the ecological functions of both pathogenic and useful bacterial isolates-information that can be used towards precautionary causes or further inquiry and useful application.

Acknowledgements
This work was supported by Alpha BioLaboratory, special thanks Dr. I-Chen for making this report possible.

References

[1] "Marine Microbes: Did You Know?" *National Oceanic and Atmospheric Administration*. n.d. Pdf. 18 July 2016.
[2] "Marine Microbes." *Australian Institute of Marine Science*. 24 July 2000. Web. 18 July 2016.
[3] "Bacteria and virus contamination of water." *AquaBright*. 2013. Web. 11 July 2016.
[4] Sharp, Merck, and Dohme Corp. "Bacterial diseases of fish: Aquarium fishes: Merck veterinary manual." *Merck Manuals*. 2009. Web. 11 July 2016.
[5] S Jeshveen, L Chai, C Pui and Son R 2012 "Optimization of multiplex PCR conditions for rapid detection of Escherichia Coli O157:H7 virulence genes". *International Food Research Journal.* **19** 461
[6] Healy B, Cooney S, O'Brien S, Iversen C, Whyte P, Nally J, Callanan J and Fanning S 2009 "Cronobacter (Enterobacter Sakazakii): an opportunistic foodborne pathogen". *Medline* **7** 339–50
[7] "Enterobacter infections: Practice essentials, background, Pathophysiology." *Medscape*. 2 June 2016. Web. 12 July 2016.
[8] Bennett H, Mok H, Gkrania-Klotsas E, Tsai I, Stanley E, Antoun N, Coghlan A, Harsha B, Traini A, Ribeiro D et al 2014 "The genome of the Sparganosis tapeworm Spirometra Erinaceieuropaei isolated from the biopsy of a migrating brain lesion" *Genome biology* **15** 510
[9] "Hydatigera taeniaeformis - WormBase paraSite." *Wormbase Parasite*. 2016. Web. 12 July 2016.
[10] WHO. "Schistosomiasis." *World Health Organization*. World Health Organization, 22 Feb. 2016. Web. 12 July 2016.
[11] Zhang H and Wang S 2013 "Rice versus Pv.: a unique pathosystem". *Current Opinion in Plant Biology* **16.2** 188–195
[12] CDC. "Diphyllobothrium - biology." *Parasites - Diphyllobothrium Infection*. CDC, 15 Aug. 2010. Web. 16 July 2016.
[13] Malcolm J, Hall N, Fung E, White O, Berriman M, Hyman R, Carlton J, Pain A, Nelson and Bowman S 2002 "Genome sequence of the human malaria parasite Plasmodium falciparum" *Nature* **419** 498-511
[14] Ni'matuzahroh, Gilewicz M, Guiliano M and Bertrand J 1999 "In-vitro study of interaction between photooxidation and biodegradation of 2-methylphenanthrene by Sphingomonas Sp. 2MPII" *Chemosphere* **38.11** 2501–07
[15] "Pseudoalteromonas." *Microbe Wiki*. n.d. Web. 12 July 2016.
[16] Comitini F and Ciani M 2009 "The zymocidial activity of Tetrapisispora Phaffii in the control of Hanseniaspora Uvarum during the early stages of winemaking." *Letters in applied microbiology* **50.1** 50–6
[17] Pfukenyi D, Mukaratirwa S, Willingham A and Monrad J 2006 "Epidemiological studies of Schistosoma Mattheei infections in cattle in the highveld and lowveld communal grazing areas of Zimbabwe." *The Onderstepoort journal of veterinary research* **73.3** 179–91
[18] Hamdan L and Fulmer P 2011 "Effects of COREXIT® EC9500A on bacteria from a beach oiled by the deepwater horizon spill" *Aquatic Microbial Ecology* **63.2** 101-09
[19] "Marinobacter hydrocarbonoclasticus." *Microbe Wiki*. n.d. Web. 12 July 2016.
[20] "Bacillus pumilus." *Microbe Wiki*. n.d. Web. 12 July 2016.
[21] Yurkov V, Jappe J and Vermeglio A 1996 "Tellurite resistance and reduction by obligately aerobic photosynthetic bacteria." *Appl Environ Microbiol* **62.11** 4195-8
[22] Chao T, Lai C and Huang J 2013 "Lactococcus Garvieae Peritoneal Dialysis Peritonitis." *Perit*

Dial Int. **33.1** 100-01
[23] "Vibrio infections: Background, Pathophysiology, Epidemiology." *Medscape*. 2 June 2016. Web. 12 July 2016.
[24] "Causes and symptoms of Vibrio parahaemolyticus - Minnesota Dept. Of health." *Minnesota Department of Health*. 10 Mar. 2015. Web. 12 July 2016.
[25] Maizels, Rick. "Nippostrongylus brasiliensis." *Rick Maizels' Group*. n.d. Web. 12 July 2016.
[26] "Providencia infections: Background, Pathophysiology, Epidemiology." *Medscape*. 10 Apr. 2016. Web. 16 July 2016
[27] Wu M, Ren Q, Durkin A, Daugherty S, Brinkac L, Dodson R, Madupu R, Sullivan S, Kolonay J and Haft D et al 2005 "Life in hot carbon monoxide: the complete genome sequence of carboxydothermus hydrogenoformans Z-2901". *PLoS Genetics* **1.5** e60
[28] "Carboxydothermus hydrogenoformans." *Microbe Wiki*. n.d. Web. 12 July 2016.
[29] Yang Y, Yang J, Wu W, Zhao J, Song Y, Gao L, Yang R and Jiang L 2005 "Biodegradation and mineralization of polystyrene by plastic-eating mealworms: part 2. role of gut microorganisms." *Environmental Science & Technology* **49.20** 12087–93
[30] Hu X, Wang C and Wang P 2015 "Optimization and characterization of biosurfactant production from Marine Vibrio Sp. Strain 3B-2." *Front Microbiol* **6** 976

IOP Conf. Series: Earth and Environmental Science **51** (2017) 012016 doi:10.1088/1755-1315/51/1/012016

Performance of the Iron-Caron Coupling Constructed Wetland for Rural Sewage Treatment

Ruize Xia, Yutao Peng, Shan Zhong, Lijun Tu, Yuanshan Xie and Lishan Zhang

Key Laboratory of Ecology of Rare and Endangered Species and Environmental Protection of Ministry Education, Guangxi Normal University, Guilin541004, P. R. China

lishan_zhang@hotmial.com

Abstract. In recent years, rural decentralized sewage treatment have gained widespread attention. Although wastewater treatment facility has been developed for rural areas, most rural population are left without adequate wastewater treatment systems. In the present study, the performance of iron-carbon coupling constructed wetland system (ICCWS) and constructed wetland system (CWS) receiving synthetic domestic wastewater were compared in side-by-side trials. Studies have found that CWS filled with spherical iron-carbon packing showed better treatment efficiency than normal CWS. When the HRT is 3 days, the ICCWS have 95.8% COD and 96.6% PO_4^{3-} removal rate ,higher than 67.0% COD and 74.3% PO_4^{3-} removal rate in CWS respectively. The use of ICCWS planted with canna and cattail proved to be efficient technology for the removal of rural wastewater pollutants.

1. Introduction

In China, rural residents accounted for 57.36% of the population. By 2014, more than 768 million people lived in the 2.79 million villages. As the living standards of the rural areas increase, rural water consumption is growing greatly ($11.6\times10^9 m^3$ per annual [1]) . As a result, excess generation of wastewater is taking place and subsequently discharged to the environment. This leads to the deterioration of water quality in rural areas. According to data released by the Ministry of Housing and Urban Rural Development of China in 2010, 94% of rural areas are left without adequate wastewater drainage and treatment systems [2]. It is a common practice to discharge the domestic wastewater without efficient and proper treatment processes to the nearby lakes or rivers, resulting in serious environmental pollution and impairment of human health [3]. Water pollutants discharged in rural areas are responsible for almost 50% of the total pollution in China. Now, septic tanks are mainly onsite treatment solution for decentralized treatment of rural domestic wastewater in China. The digested wastes generated in septic tanks are commonly used in agriculture as fertilizers or discharged to nearby lakes or rivers. Both the disposing ways could causing potential environmental risks. Excessive application of digested wastes in farmland for a long time can cause groundwater pollution and the point source and the non-point source pollutions to varying degrees [4]. There is a pressing need to develop dependable technologies that are of simple structure, low cost, low maintenance and highly efficient for rural decentralized wastewater treatment.

Conventional centralized treatment systems are operated in large and small cities with aim to control water environment pollution. But the centralized treatment systems are impractical in rural areas due to the requirement of considerable investment in construction and operation. It's particularly urgent for the small towns and rural areas in our country to develop a simple, reliable and affordable wastewater treatment solution. The use of constructed wetlands have turn out to be the practical way. Constructed wetlands work as a typical artificial ecological treatment system and have been commonly used in many small towns and rural areas because of their advantages like low investment, low energy consumption and convenient operation management [5].

The key elements of the applications of constructed wetlands in sewage treatment are plants and media [6-7]. The traditional vertical flow constructed wetland technology had been studied by some researchers, but the process about Iron-Carbon coupling Constructed Wetland System (ICCWS) has been not reported yet. So in this work, an attempt has been made to integrate iron carbon micro-electrolysis technology with constructed wetland to treat rural domestic wastewater. ICCWS offer several advantages such as "using waste to treat waste", high performance efficiency and low investment. The system can be widely applied. The iron-carbon reduction method had been studied by many researchers. Iron-carbon micro-electrolysis process intergraded with biochemical treatment process have several uniqueness including increase biodegradability of the wastewater, improvement of the wastewater quality and decrease the biochemical treatment load. The iron-carbon micro-electrolysis has an obvious advantage in improving the wastewater treatment ability. Given the treatment of high strength organic wastewater, there are several solutions integrated with micro-electrolysis like biological contact oxidation process integrated with micro-electrolysis system, biological contact oxidation process integrated with anaerobic digestion and micro-electrolysis system and SBR integrated with micro-electrolysis system [8-9]. Under anoxic conditions, the micro-electrolysis treatment operates on a principle of galvanic cell reaction. Anodic Fe(0) serves as an active electron donor ($E^0(Fe^{2+}/Fe^0) = -0.44$ V). Hydrogen ions (H^+) was first reduced into hydrogen (H_2) by accepting electrons, and then the freshly formed hydrogen drove the reductions in cathodic region [10-11].

Anode: $Fe - 2e \rightarrow Fe^{2+}$

Cathode: $2H^+ + 2e \rightarrow 2[H] \rightarrow H_2$

Products released from the galvanic cell reaction include hydroxyl, atomic hydrogen and Fe(II) which have high activities to decompose contaminants. The structure of the organic compounds in wastewater can be destroyed, thus, improving the sewage biodegradability and decreasing the COD concentration of the effluent [12-13]. In addition, Fe(II) is often used as a coagulant to remove phosphate in wastewater. Thus, it could be promising to combine micro-electrolysis with CWS together in a same system to enhance nutrient removal efficiency [14-15].

According to the government report [1], the wastewater generated in rural areas mainly comprises of grey water from household activities, toilet wastewater and livestock wastewater from family backyard. The wastewater typically contains no heavy metals,

CODcr(150-400mg/L) and TP(2.0-6.0mg/L) [16]. Dispersed distribution of the population in rural areas determines the inherent variability of rural domestic wastewater characteristics like quantity and quality. Additionally, Rural wastewater treatment emphasizes on removal of the main pollutant of the CODcr and Phosphorus [17]. Ideally, a practical decentralized wastewater treatment systems can fill the criteria of affordability, sustainability and environmental friendly. The designed system is to approach a high removal efficiency of the COD and Phosphorus at reduced treatment costs through the combination of micro-electrolysis with constructed wetland. A new packing material named as spherical iron-carbon packing composes of one of bed medias of constructed wetland for COD reduction and phosphorous retention. In this study, laboratory scale vertical subsurface constructed wetlands integrated with spherical iron-carbon packing were set up to evaluate and compare the removal efficiency of COD and PO_4^{3-} in treating synthetic domestic wastewater. The study was also conducted to evaluate the stability of the system performance under different experiment conditions.

2. Methods

2.1. Test materials and devices

Two types of laboratory scale reactors were set up using plexiglass containers (height, 80cm, diameter, 25cm). Figure 1 represents the schematic diagram of the designed traditional CWS set-up with 5~8 mm diameter gravel. Figure 2 represents the schematic diagram of the designed ICCWS set-up with 10 mm diameter iron-carbon packing on bottom and 5~8 mm diameter gravel on top. Wastewater was pump into vale 1 via a peristaltic pump and discharged from valve 3. There were various aquatic plants like cattail, Cyperus alternifolius, canna on the wetland top layer. The CWS and ICCWS were operated under the same conditions and the same time, to ensure the comparability of the research systems.

2.2. Test methods and water quality

The synthetic rural domestic wastewater was prepared by dissolved KH_2PO_4, glucose($C_6H_{12}O_6$), beef extract, peptone and starch in tap water to make it contain about 3~5mg/L PO_4^{3-} and 200~230mg/L COD. The hydraulic retention time (HRT) in this experiment was set to a 2 and 3 day period. The COD and PO_4^{3-} of influent and effluent water was measured in consecutive time, and their removal efficiencies finally calculated. The water quality of the influent water, effluent water and water samples respectively from various sampling points was analyzed, using measuring methods in accordance with the 2002 "Water and Wastewater Monitoring and Analysis Methods" released by the Ministry of Environmental Protection of the People's Republic of China (MEP), formerly State Environmental Protection Administration (SEPA). The COD was measured using the potassium dichromate method-UV spectrophotometric method [18-19], and PO_4^{3-} using the ammonium molybdate spectrophotometric method [20].

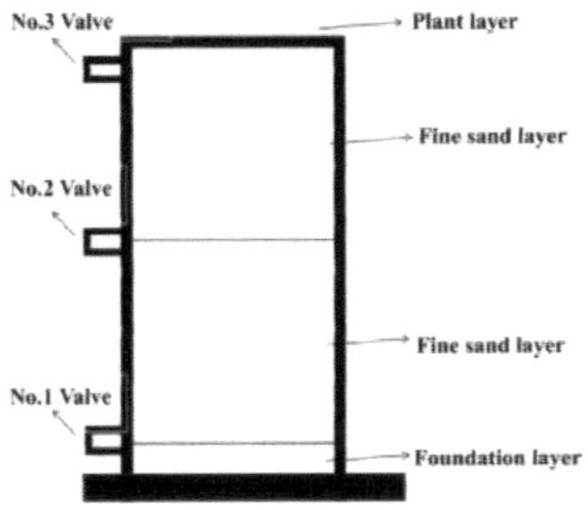

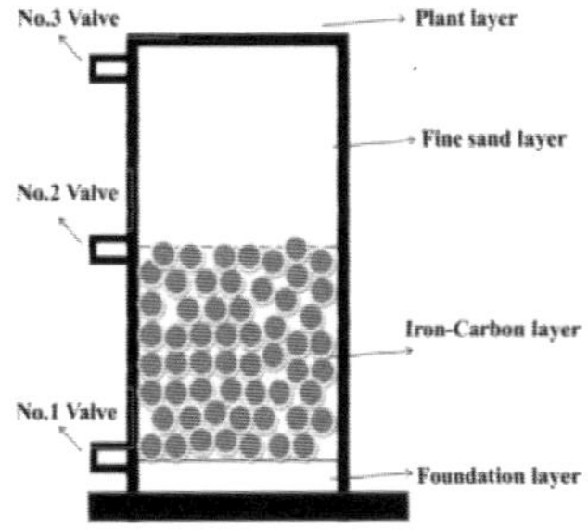

Figure 1. Traditional CWS cylindrical simulation diagram.

Figure 2. The new ICCWS cylindrical simulation diagram.

3. Test results and discussion

3.1. Removal of COD

It is known that different medias in constructed wetland play a certain role in COD removal efficiency. In addition to absorb some inorganic pollutants, the media providing a larger surface area can create good living conditions for the reproduction and role of microorganisms. Therefore, more efficient purification can be achieved through the effective adsorption and degradation of water pollutants in the wastewater. According to the figures 3-5 below, the high removal values of COD may be explained as in ICCWS, settable organics were rapidly removed by deposition and filtration due to coagulation effection of Fe^{2+}, while the organic compounds were degraded by the micro-electrolysis function.

3.2. Removal of phosphorus

The removal of phosphorus by constructed wetlands mainly depends on the filler layer's substrate, the aquatic plants and microorganisms in constructed wetland, as well as the combined effects of the three. The canna-ICCWS and cattail-ICCWS exhibited excellent performance in PO_4^{3-} removal throughout the whole study period(figures 6-8). The PO_4^{3-} removal efficiencies of the canna-ICCWS and cattail-ICCWS grow with the increase of HRT 2 to 3days. The filler plays the most important role in Phosphorus removal, with the combined effect of aquatic plants and microorganisms being second only after the filler substrate. Fe^{2+} released from the galvanic reaction of iron carbon granular filler layer form iron phosphate precipitates with PO_4^{3-} removal in wastewater, thus improving PO_4^{3-} removal efficiency.

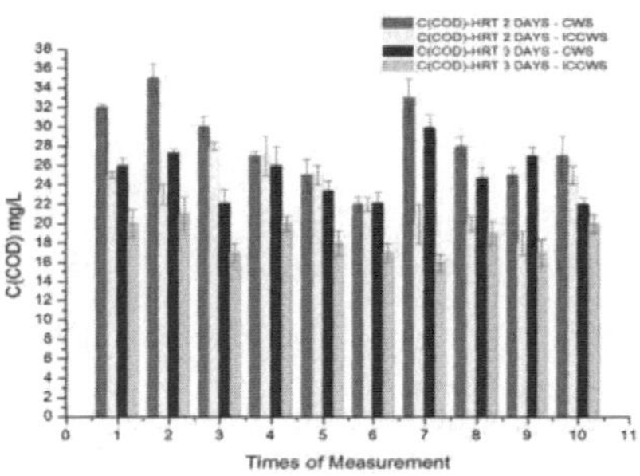

Figure 3. The Canna-CWS/ICCWS .in Different HRT.

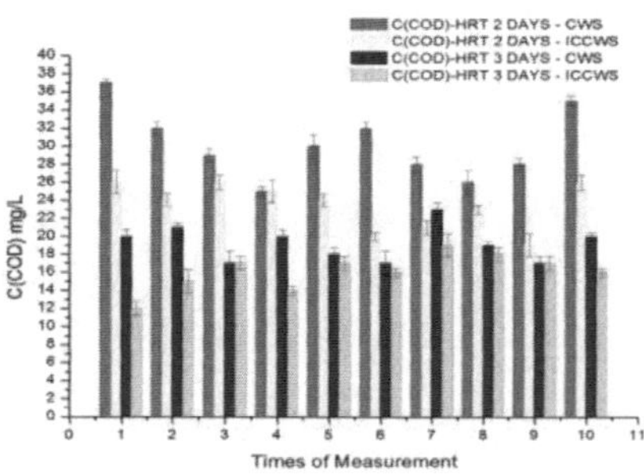

Figure 4. The Cattail-CWS/ICCWS in different HRT.

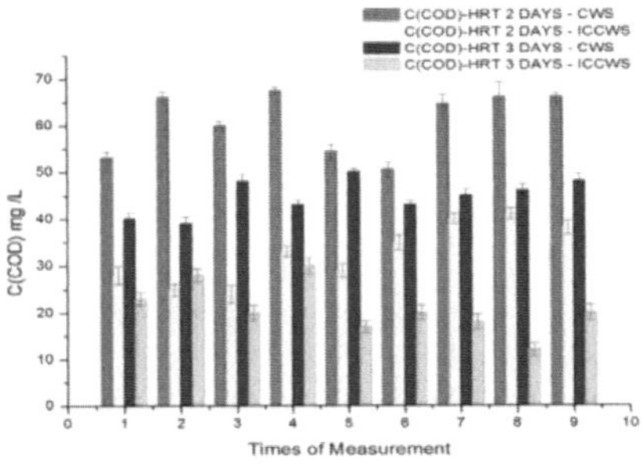

Figure 5. The Cyperus alternifoliu s-CWS/ICCWS in different HRT..

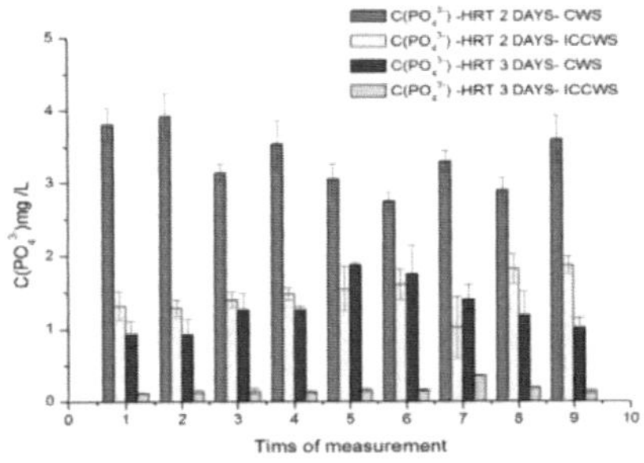

Figure 6. The Canna-CWS/ICCWS in different HRT.

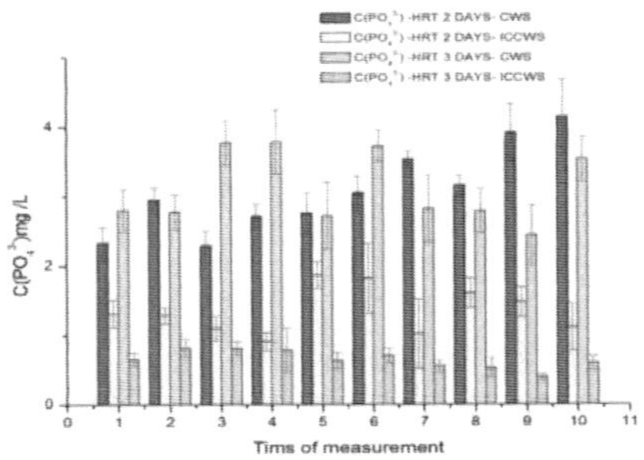

Figure 7. The Cattail-CWS/ICCWS. in different HRT.

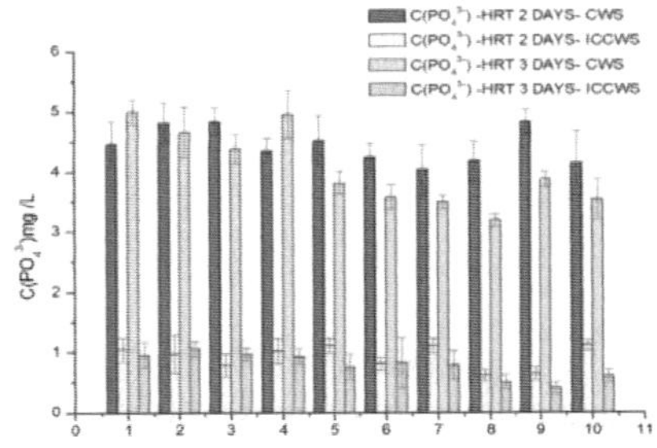

Figure 8. The Cyperus alternifolius-CWS/ICCWS in different HRT .

3.3. The comparison of different plants

When canna was planted in the both systems to treat the rural sewage. According to the figures 9-10 below, the resulting mean COD removal efficiency of 87% and 88%, respectively with corresponding HRT of 2 and 3 days, observed in the CWS laboratory system, are in contrast to the higher COD removal of approximately 87% and 91% in the ICCWS. When cattail was select as the wetland plant, the average effluent COD concentrations from CWS reach 30.2mg/L and 23.4mg/L, respectively with corresponding HRT of 2 and 3 days. In comparison, the average effluent from ICCWS has lower COD concentrations of 19.2mg/L and 16.1mg/L. When Cyperus alternifolius was select as the wetland plant, the average effluent COD concentrations from ICCWS reach 66.0mg/L and 8.33mg/L, respectively with corresponding HRT of 2 and 3 days. In comparison, the average effluent from CWS has higher COD concentrations of 80.33 mg/L and 31.8mg/L.

The results indicate that the COD removal efficiency of the ICCWS is better than that of the CWS. With the increase in HRT, the COD removal efficiency increases. Among the three plant species, the COD removal efficiency of ICCWS with canna and cattail are best.

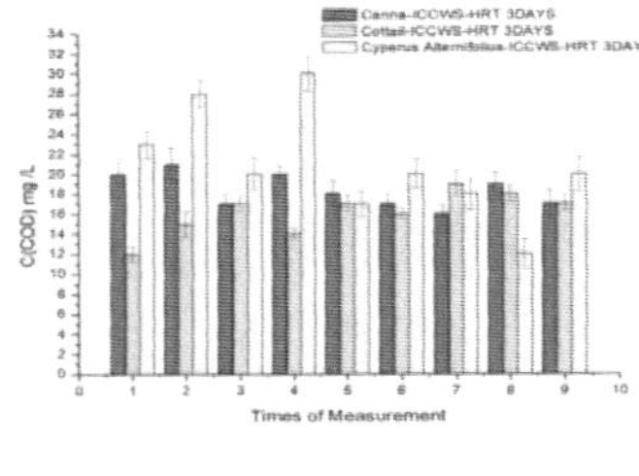

Figure 9. The COD of ICCWS in different plants.

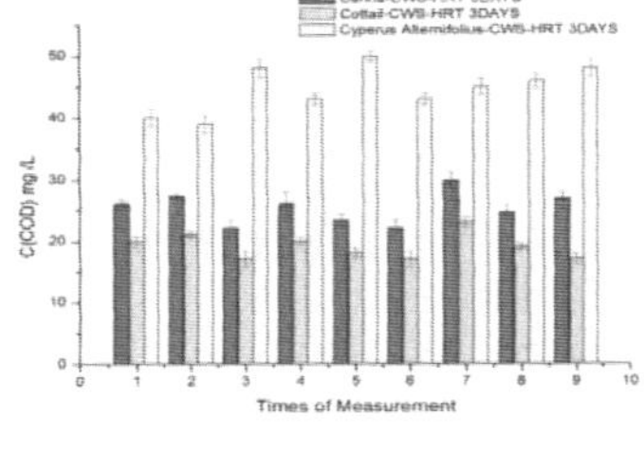

Figure 10. The COD of CWS in different plants

4. Then main research and conclusion

The use of ICCWS planted with canna and cattail proved to be efficient technology for the removal of rural wastewater pollutants. The system can effectively degrade COD and phosphorus in rural household wastewater. The effluent water quality can meet the first class of A standard of the "Discharge standard of pollutants for municipal wastewater treatment plant" (GB18918-2002). The process is also very practical when it is used to deal with household wastewater in tourist areas, resorts or scenic areas without service of urban drainage system. The treated water can be recycled as irrigation water to generate economic benefits, further reducing the cost of investment.

Acknowledgements

The study was supported by the National Natural Science Foundation of China (No. 51469006), Guangxi Natural Science Foundation (2014GXNSFAA118298), Guangxi Science and Technology Council Funds (1598016-4), Guilin Science and Technology Council Funds (20130113-3) and Key Laboratory of Ecology of Rare and Endangered Species and Environmental Protection (Guangxi Normal University), Ministry of Education, China.

References

[1] Ministry of Housing and Urban Rural Development of China 2014 *Chinese Urbanrural Construction Statistical Yearbook* pp 133–135.
[2] Guo X S, Liu Z H, Chen M X, Liu J X, Yang M 2014 *Higher Education Press.* **8** 929
[3] Li C, Yan Z J, Zhang J W *et al.* 2014 *J. Agri. Environ. Sci.* **23** 1560
[4] Yan Z J, Chen S, W M F *et al.* 2015 *J. Agri. Res. Environ.* **123** 31
[5] Lu S Y, Jin X C *et al.* 2006 *Acta Ecol. Sin.* **14** 2670
[6] A V Dordio, A J P Carvalho 2013 *J. Hazard.Mater.* **15** 272
[7] Wu H M, Zhang J 2015 *Bioresour. Technol.* **175** 594
[8] Zhang Z S,Yang Y F ,Gong-Di XU, Chen JH 2008 *Pollution Control Tech.* **51** 54
[9] Qin i, Li Y, Jin L *et al.* 2006. *Jiangsu Environ. Sci. Tech.* **23** 46
[10] Yuan S B 2015 *Iron carbon micro electrolysis pretreatment pharmaceutical wastewater* (Wuhan: Wuhan University of Science and Technology)p 53
[11] Yang L, Song X, Wang Y J *et al.* 2013 *J. Environ. Eng.* **23** 268
[12] Yu X, Lin G 2011 *Industrial Water Treatment* **65** 30
[13] Han W Q, Xue H M, Chen Y *et al.* 2011 *Environmental engineering* **78** 45
[14] Si G L, Liu J H, Yu D Q 2013 *The academic annual meeting of China environmental science society* pp 325–329
[15] Geng S Q, Liu H G, Luo W *et al.* 2012 *J. Environ. Eng.* **56** 242
[16] Ma D D, Wen C, Ji M 2013 *Chemical progress* **23** 205
[17] Liang Z, Liu W *et al.* 2009 *Journal of Anhui agriculture*. **34** 78
[18] Shi C;Zhang J;Liu C;Li Z X;Wu G;Yang J L. 2016 *Journal of Hebei University of Science & Technology* **37** 1
[19] Wei L X, He Y, Li C H. 2004 *Environmental science and technology* **26** 33
[20] The state environmental protection administration 2012 *Ammonium molybdate spectrophotometric method* (Water and wastewater monitoring analysis method, Chemical industry press)

Evaluation and Prediction of Water Resources Based on AHP

Shuai Li[1,2] **and Anqi Sun**[1,3]

[1] College of Communication Engineering, Jilin University, Changchun, China 130012

[2] luckyappleLS@163.com [3] 18166880717@163.com

Abstract. Nowadays, the shortage of water resources is a threat to us. In order to solve the problem of water resources restricted by varieties of factors, this paper establishes a water resources evaluation index model (WREI), which adopts the fuzzy comprehensive evaluation (FCE) based on analytic hierarchy process (AHP) algorithm.
After considering influencing factors of water resources, we ignore secondary factors and then hierarchical approach the main factors according to the class, set up a three-layer structure. The top floor is for WREI. Using analytic hierarchy process (AHP) to determine weight first, and then use fuzzy judgment to judge target, so the comprehensive use of the two algorithms reduce the subjective influence of AHP and overcome the disadvantages of multi-level evaluation. To prove the model, we choose India as a target region. On the basis of water resources evaluation index model, we use Matlab and combine grey prediction with linear prediction to discuss the ability to provide clean water in India and the trend of India's water resources changing in the next 15 years. The model with theoretical support and practical significance will be of great help to provide reliable data support and reference for us to get plans to improve water quality.

1. Introduction

1.1. Background

Water shortage has become a global problem to be solved, more and more countries and regions has been caught in moderate or severe water shortages. In the 21st century, the scarcity of water resources is becoming a kind of precious resources, water resources is not only a resource problem, more be related to the national economy, sustainable development and social stability and security of major strategic issues. Therefore, when the human develop the water resources, must according to the law of the water cycle, and make full use of water resources reasonably.

1.2. Conspectus

We need a model to measure the ability of a region to provide clean water, the mathematical model can be transformed into a model of water resources evaluation index. We set up the index as the WREI (Water resources evaluation index), which is used to judge if the region clean water is abundant. Considering there are many factors and the influences of water resources are different, we use the analytic hierarchy process (AHP) to help the establishment of the model. We are going to regard WREI as the target layer, considering the water resource, population and water use and capacity will determine the final value of the WREI, so put the three index as criterion layer. And all details of factors as the bottom layer .Considering when using the simple analytic hierarchy model to judge the water abundance degree, larger proportion of subjective factors may cause large deviation, so we can use fuzzy comprehensive evaluation (FCE) method to remove subjective views [1]. Through the multilayer analysis method to determine the weights, and then through the fuzzy comprehensive evaluation to evaluate target, Combining with the two models will reduce the subjective influence of repeatedly using AHP, and overcome the disadvantages of multi-level evaluation.

The aim of the model is to get the final equation of water evaluation, determine how various factors affect the value of their weights of evaluation index. The conclusion of this model can be used for a particular area, water design intervention programs to change the value of the factors that tend to better water resources evaluation index value, which is to improve the water situation.

2. Model establishment and implementation

2.1. Model assumptions

- Ignore the secondary factors and the influence factors of water resources is only related to the main factors
- The search data on www.fao.org/nr/water/aquastat/water_res/index.stm is the theoretical data
- Per Capita GDP is roughly on behalf of the economic ability of regional water purification
- A region's supplement and demand of water resources distribute equally, i.e., ignoring the allocation of resources

2.2. Symbol definition

- WREI :water resources evaluation index, the target value in the model
- u_i :on behalf of the elements at all levels
- W_n :the weight matrix of target element n
- A:weight matrix
- R:membership matrix

2.3. Establishment of an AHP Model

The model is judged by water resources, population and water use and capacity [2].They are provided as u1,u2 and u3.Water resources depend on national rainfall index, internal renewable water resources and External renewable water resources called as u11, u12 and u13 respectively as shown in Figure1. Similarly, we also let u2 and u3 be stratified.

2.3.1. Data stratification. WREI as the target layer, criterion layer includes three factors, namely the WSI = {u1, u2, u3}, solution layer includes seven factors, namely the u1 = {u11, u12, u13}, u2 = {u21, u22}, u3 = {u31, u32}, the relation between each factor hierarchy is shown in Figure 1.

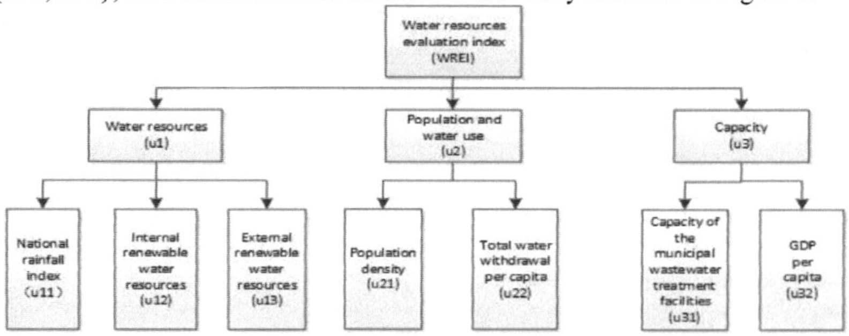

Figure 1. Hierarchy diagram

The WREI and index value range from 0 to 1; the bigger the index value is, the better water security situation will be. In general, the mathematical expression of water security index is

$$WREI = \sum_{i=1}^{n} w_{x,i} X \quad (1)$$

WREI refers to the water abundance index value of a certain region; W refers to the weight of sub-index X of the WREI.

2.3.2. Analytic hierarchy process. We use Satty's 1-9 Scale method to make thinking quantitative [3].Then we determine the index weight and have a consistency check.

We use water resources, population and water use, capacity in the second layer to solve three weights as an example.

2.3.2.1. Determination of judgment matrix derived from Satty's 1-9 Scale method: $P=\begin{pmatrix} 1 & 1 & 7 \\ 1 & 1 & 7 \\ \frac{1}{7} & \frac{1}{7} & 1 \end{pmatrix}$

2.3.2.2. Feature vector solution. To solve the characteristic vector W_{WREI}, according to the square root method [4].

1) Calculating the elements' product of each line in judgment matrix P

$$P(1)=1*1*7=7; P(2)=1*1*7=7; P(3)=\frac{1}{7}*\frac{1}{7}*1=\frac{1}{49};$$

2) Calculate the three square root W_i of P (I): $W_1=1.9129; W_2=1.9129; W_3=0.2733;$

3) Do normalization and regularization processing for W = (W1, W2, W3),

$$W_i'=\frac{W_i}{\sum_{i=1}^{3} W_i} \quad (2)$$

Get: $W_1'=0.4667, W_2'=0.4667, W_3'=0.0666$

The feature vector: $W_{WREI}=(0.4667,0.4667,0.0666)$

2.3.2.3. Consistency check. We will need a consistency check of judgment matrix to test above characteristic vector weights allocation is reasonable, the method is as follows:

1) Calculate the maximum eigenvalue λ_{max} of judgment matrix

$$\lambda_{max}=\sum_{i=1}^{3}\frac{(PW_{WREI})_i}{nW_i'}=\frac{1}{n}\sum_{i=1}^{3}\frac{(PW_{WREI})_i}{W_i'} \quad (3)$$

(n refers to the amount of transversal vector)

In the formula, $(PW_{WREI})_i$ represent the element i of PW_{WREI} ,and n=3

$$PW_{WREI}=\begin{pmatrix} (PW_{WREI})_1 \\ (PW_{WREI})_2 \\ (PW_{WREI})_3 \end{pmatrix}=\begin{pmatrix} 1 & 1 & 7 \\ 1 & 1 & 7 \\ \frac{1}{7} & \frac{1}{7} & 1 \end{pmatrix}\begin{pmatrix} 0.4667 \\ 0.4667 \\ 0.0666 \end{pmatrix}, \text{ Get: } \lambda_{max}=3$$

2) Consistency check, use the test formula:

$$CR = \frac{CI}{RI}, CI = \frac{\lambda_{max} - n}{n-1} \tag{4}$$

With n = 3, RI = 0.58, $\lambda_{max} = 3$, get: CR = 0.7 < 0.1

Show a satisfactory consistency judgment matrix P, so each component $W_{WREI} = (0.4667, 0.4667, 0.0666)$ can be used as a weight.

2.3.2.4. Determination of the weights of every index. According to the above method, can calculate the indexes' judgment matrix , weight and consistency check results as Table 1,Table 2,Table 3,Table 4, you can see the consistency of judgment matrix is satisfied, illustrating the effectiveness of the judgment matrix of each index.

Table 1. Judgment matrix WREI_u1, $W_{WREI\_u1} = \{0.7778, 0.1111, 0.1111\}$

u1	u11	u12	u13	Weight WWREI_u1	Consistency check
u11	1	7	7	0.7778	λmax=3 CI=0 CR=0<0.1
u12	1/7	1	1	0.1111	
u13	1/7	1	1	0.1111	

Table2. Judgment matrix WREI_u2

$W_{WREI\_u2} = \{0.8333, 0.1667\}$

u21	u22	Weight WWREI_u2	Consistency check
1	5	0.8333	λmax=2 CI=0 CR=0<0.1
1/5	1	0.1667	

Table3. Judgment matrix WREI_u3

$W_{WREI\_u3} = \{0.8571, 0.1429\}$

u3	u31	u32	Weight WWREI_u3	Consistency check
u31	1	6	0.8571	λmax=2 CI=0 CR=0<0.1
u32	1/6	1	0.1429	

WSI	u1	u2	u3	Weight WWREI	Consistency check
u1	1	1	7	0.4667	λmax=3 CI=0 CR=0<0.1
u2	1	1	7	0.4667	
u3	1/7	1/7	1	0.0666	

Table 4. Judgment matrix WREI, $W_{WREI} = \{0.4667, 0.4667, 0.0666\}$

2.4. Establishment of a fuzzy comprehensive evaluation model (FCE)

2.4.1. Principle of a fuzzy evaluation model. Fuzzy evaluation factors is mainly composed of matrix of relation, factor sets and judgment sets.

1) Factor sets: $U = \{U_1, U_2, U_3, ..., U_k\}$ $(k = 1,2,3...)$;

2) Judgment sets: $V = \{V_1, V_2, V_3, ..., V_k\}$ $(k = 1,2,3...)$;

Fuzzy rating constitute a set V, and the number should be same as the number of levels of factors' division.

2.4.2. The single factor evaluation. R refers to the single factor evaluation matrix: $R = \begin{pmatrix} r_{11} & r_{12} & \cdots & r_{1m} \\ r_{21} & r_{22} & \cdots & r_{2m} \\ \cdots & \cdots & \cdots & \cdots \\ r_{n1} & r_{n2} & \cdots & r_{nm} \end{pmatrix}$

2.4.3. Comprehensive evaluation. Choose the fuzzy matching model M, for weight $A = (a_1, a_2, a_3, \ldots, a_j)\ (j = 1,2,3, \ldots)$

According to the actual data situation, determine the plus or minus characteristic of elements in weight A and get the weight A, and then we can get comprehensive evaluation.

$$WREI = A' \cdot R = (wrei_1, wrei_2, wrei_3, \ldots, wrei_m) \quad (5)$$

The bigger WREI value is, the greater the regional water security situation will be, that is, the region can provide more clean water.

3. Model validation and prediction

We choose India as an example of explanation because it is known that the water resources was of severe or moderate shortage in most regions of India. We will use the data of India to prove the AHP-FCE Model and predict the water resources evaluation index in future.

3.1. Data and Data Processing

3.1.1. Data. We found and organized the required data on the reference website. Because the model is of fuzzy evaluation based on analytic hierarchy process (AHP), according to the water resources, population and water use, capacity, we divide the data into three parts.

We take part 1 as an example. We list the survey data about water resources on www.fao.org/nr/water/aquastat/water_res/index.stm in Table 5 as follow:

Table 5. Data of Water resources

(mm/year)/(10^9 m3/year)/(10^9 m3/year)

Factors / Years	National Rainfall Index	Internal renewable water resources	External renewable water resources
1958-1962	1703	1446	464.9
1963-1967	1704	1447	465.9
1968-1972	1705	1448	466.9
1973-1977	1706	1449	467.9
1978-1982	1707	1450	468.9
1983-1987	1708	1451	469.9
1988-1992	1709	1452	470.9
1993-1997	1710	1453	471.9
1998-2002	1711	1454	472.9
2003-2007	1712	1455	473.9
2008-2012	1713	1456	474.9
2013-2017	1714	1457	475.9

3.1.2. Data processing to get WREI. (Table 5 is an example for complete process, the other two parts are in the same way.)

We need to normalize the element of each year in the table, because the way to deal with these three tables is same. We take the Table 5 as an example to write the specific process of solution W_{WREI_u1}.

1） Normalize each element in Table 5， then we can get Table 6 as follow:

Table 6. Normalized matrix of Water resources

(mm/year)/(10^9 m3/year)/(10^9 m3/year)

Years \ Factors	National rainfall index	Internal renewable water resources	External renewable water resources
1958-1962	0.471236061	0.400121752	0.128642187
1963-1967	0.471121679	0.400066355	0.128811966
1968-1972	0.471007486	0.40001105	0.128981464
1973-1977	0.470893483	0.399955836	0.12915068
1978-1982	0.470779668	0.399900714	0.129319617
1983-1987	0.470666042	0.399845683	0.129488275
1988-1992	0.470552603	0.399790743	0.129656654
1993-1997	0.470439352	0.399735894	0.129824754
1998-2002	0.470326287	0.399681135	0.129992578
2003-2007	0.470213409	0.399626466	0.130160125
2008-2012	0.470100716	0.399571887	0.130327396
2013-2017	0.469988209	0.399517398	0.130494392

2) After transferring them, we get the single factor evaluation matrix:

$$R_{WREI\_u1} = \begin{pmatrix} 0.4712 & 0.4711 & 0.4710 & 0.4709 & 0.4708 & 0.4707 & 0.4706 & 0.4704 & 0.4703 & 0.4702 & 0.4701 & 0.4700 \\ 0.4001 & 0.4001 & 0.4000 & 0.3999 & 0.3998 & 0.3998 & 0.3998 & 0.3997 & 0.3997 & 0.3996 & 0.3996 & 0.3995 \\ 0.1286 & 0.1288 & 0.1290 & 0.1292 & 0.1293 & 0.1295 & 0.1297 & 0.1298 & 0.1300 & 0.1302 & 0.1301 & 0.1305 \end{pmatrix}$$

3) Weight table available from before: $W_{WREI\_u1} = (0.7778, 0.1111, 0.1111)$

4) For the choice of the fuzzy model, we think a balanced consideration of weights in all factors, so we choose the weighted average model, namely, using the fuzzy algorithm formula for judging value of $WREI_{u1}$ in each time period:

$$WREI_{u1} = W_{WREI\_u1} \cdot R_{WREI\_u1} \tag{6}$$

Using Matlab to solve more complex matrix operations, and the result is：

$WREI_{u1} = (0.4252\ 0.4252\ 0.4251\ 0.4251\ 0.4250\ 0.4249\ 0.4249\ 0.4247\ 0.4246\ 0.4246\ 0.4245\ 0.4244)$

As the same theory, we can see single factor evaluation matrix of u2 and u3 are respectively

$$R_{WREI\_u2} = \begin{pmatrix} 0.1927 & 0.2117 & 0.2262 & 0.2538 & 0.2708 & 0.2895 & 0.3319 & 0.3322 & 0.3662 & 0.3689 & 0.3794 & 0.4067 \\ 0.8073 & 0.7883 & 0.7738 & 0.7462 & 0.7292 & 0.7105 & 0.6681 & 0.6678 & 0.6338 & 0.6311 & 0.6206 & 0.5933 \end{pmatrix}$$

$$R_{WREI\_u3} = \begin{pmatrix} 0.0001 & 0.0004 & 0.0012 & 0.0013 & 0.0014 & 0.0017 & 0.0038 & 0.0047 & 0.0046 & 0.0034 & 0.0031 & 0.0034 \\ 0.9999 & 0.9996 & 0.9988 & 0.9987 & 0.9986 & 0.9983 & 0.9962 & 0.9953 & 0.9954 & 0.9966 & 0.9969 & 0.9966 \end{pmatrix}$$

The weight table of u2 and u3 are respectively

$$W_{WREI\_u2} = \{0.8333, 0.1667\}\ ;\ W_{WREI\_u3} = \{0.8571, 0.1429\}$$

Using the fuzzy algorithm formula for judging value of $WREI_{u2}$, $WREI_{u3}$ in each time period:

$WREI_{u2} = (0.2952\ 0.3078\ 0.3175\ 0.3359\ 0.3472\ 0.3597\ 0.3879\ 0.3881\ 0.4108\ 0.4126\ 0.4196\ 0.4378)$

$WREI_{u3} = (0.1430\ 0.1432\ 0.1438\ 0.1438\ 0.1439\ 0.1441\ 0.1456\ 0.1463\ 0.1462\ 0.1453\ 0.1451\ 0.1453)$

5) Make a comprehensive evaluation of factor sets

$U_{WREI} = \{U_{WREI\_u1}, U_{WREI\_u2}, U_{WREI\_u3}\}$, so as to get the following data in Table 7:

Table 7. Data of WREI

Factors / Years	Water resources (WREIu1)	Population and water use (WREIu2)	Capacity (WREIu3)
1958-1962	0.4252	0.2952	0.143
1963-1967	0.4252	0.3078	0.1432
1968-1972	0.4251	0.3175	0.1438
1973-1977	0.4251	0.3175	0.1438
1978-1982	0.425	0.3472	0.1439
1983-1987	0.4249	0.3597	0.1441
1988-1992	0.4249	0.3597	0.1456
1993-1997	0.4247	0.3881	0.1463
1998-2002	0.4246	0.4108	0.1462
2003-2007	0.4246	0.4126	0.1453
2008-2012	0.4245	0.4126	0.1451
2013-2017	0.4244	0.4378	0.1453

6) Normalize each element for Table 7.

7) After transferring them, we get the single factor evaluation matrix:

$$R = \begin{pmatrix} 0.4925 & 0.4853 & 0.4796 & 0.4796 & 0.4639 & 0.4575 & 0.4568 & 0.4428 & 0.4326 & 0.4322 & 0.4322 & 0.4212 \\ 0.3419 & 0.3513 & 0.3582 & 0.3582 & 0.3790 & 0.3873 & 0.3867 & 0.4047 & 0.4185 & 0.4199 & 0.4201 & 0.4345 \\ 0.1656 & 0.1634 & 0.1622 & 0.1622 & 0.1571 & 0.1552 & 0.1565 & 0.1525 & 0.1489 & 0.1479 & 0.1477 & 0.1442 \end{pmatrix}$$

8) Weight table available from before: $W_{WREI} = (0.4667, 0.4667, 0.0666)$

9) Considering the water resources and the capacity are the positive effects of water resources while population and water use is the opposite effects of water resources, so we get the real weight matrix: $W_{WREI'} = (0.4667, -0.4667, 0.0666)$

The evaluation of estimate of each factors of WREI in the period

$$WREI = W_{WREI}' \cdot R \tag{7}$$

So the value of WREI is

$WREI = (0.0813, 0.0734, 0.0675, 0.0675, 00501, 0.0431, 0.0431, 0.0279, 0.0165, 0.0156, 0.0155, 00034)$

3.2. Grey prediction

Substitute the value of WREI into the grey forecasting model, we can get accumulated value formula of once fitting formulax1(t+1)=-0.49986exp(-0.17854t)+(0.58116) and accumulated value formula of twice fitting formula x1(t+1)=-0.51421exp(-0.17854t)+(0.59075), and their prediction results of WREI value is as follows in Figure 2 and Figure 3:

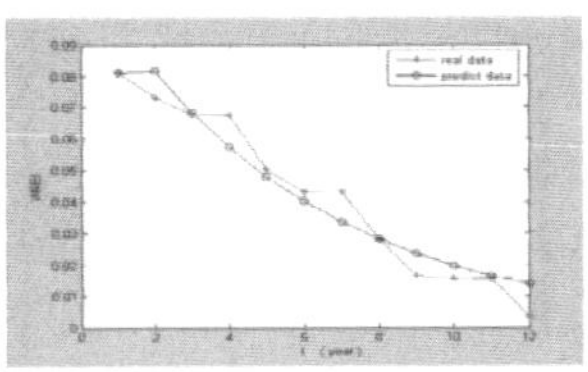

Figure 2. Once fitting prediction results

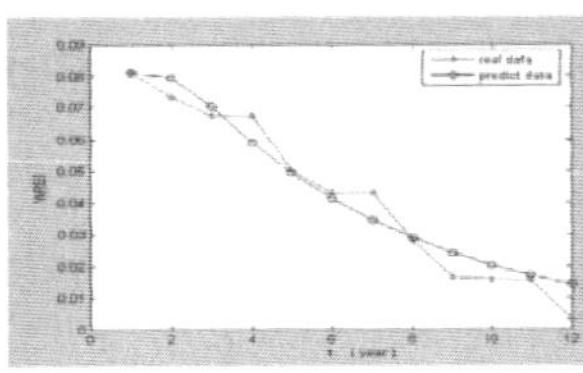

Figure 3. Twice fitting prediction results

3.3. Improvement of Grey Prediction for Linear Prediction

3.3.1. Problem analysis. By above Figure 2 and Figure 3, although the predicted values of the grey prediction is relatively close to the actual value, but in the process of variable t increase, predicted values will tend to be 0, which doesn't conform to the actual shortage situation, so if we chose linear prediction model [5], the effect will be better. And then we get the following equation and Figure 4 by Matlab.

WREI=-8315950251235889/1152921504606846976*x+3205089029355199/3602879

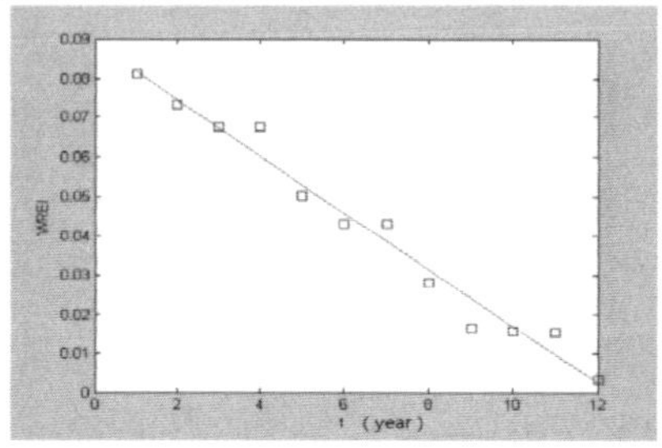

Figure 4. Linear prediction results

3.3.2. *Data and image of the predicted results.* According to the relational expression, I get the predictions in the next 15 years (2016-2030) of water situation:-0.0048 -0.0120 -0.0192 -0.0264 -0.0337 -0.0409 -0.0481 -0.0553 -0.0625 -0.0697 -0.0769 -0.0842 -0.0914 -0.0986 -0.1058

3.3.3. Prediction conclusion. Obviously, according to the prediction of water resources, short supply will occur in 2016.That is to say India's water consumption is greater than the sum of natural water and the water provided by human intervention. The situation becomes increasingly serious in the future, so if there are no special reasons within the next 15 years, India's water resources will become increasingly scarce, bringing significant pressure to economic development and people's life.

4. Conclusion

In this paper, we established the model of fuzzy comprehensive evaluation based on AHP to evaluate regional capability of providing clean water. Under the condition of the support of reliable data, we can calculate WREI, the greater the value is, the more abundant water resources are. By fitting analysis of WREI, we can get the trend of water resources in the next few years. The change factor value by having a plan can ultimately increase the value of WREI, namely to improve water quality

Despite our evaluation model include several layers, the modelling domain presented by this problem is vast, and there is a large amount of room for improvement. We believe that the work we have presented here is a significant and successful attempt at solving this problem.

5. Reference

[1] Jungang Luo 2008 Based on the entropy weight fuzzy comprehensive evaluation model of water shortage risk and applications *Journal of hydraulic engineering* **39** 1092

[2] Rong Wu 2013 The water security of fuzzy comprehensive evaluation model based on analytic hierarchy process (AHP) and its application *Journal of water resources and water engineering* **24** 139

[3] Xunlian Si, Jinxian Chen and Xinyi Xu 2006 The establishment of the water level pricing model *Journal of management engineering* **20** 74

[4] Hongrui Wang, Longxia Qian and Xinyi Xu 2009 Based on the fuzzy probability of the water shortage risk assessment model and its application *Journal of hydraulic engineering* **40** 813

[5] Xiaofang Shen 2013 Water resources risk prediction research *Journal of Yangtze university* **10** 44

IOP Conf. Series: Earth and Environmental Science 51 (2017) 012018 doi:10.1088/1755-1315/51/1/(

Performance of Multilevel Contact Oxidation in the Treatment of Wastewater from Automobile Painting Industry

Tong Zhu[1], Yufang Zhu[1], Udo Fienko[2], Xie Yuanhua[1] and Zhang Kuo[1]

[1]School of Mechanical Engineering & Automation, Northeastern University, Shenyang 110819, China.
[2]BMW Brilliance Automotive Ltd., Shenyang 100044, China

Prof. Zhu Tong, E-mail: tongzhu@mail.neu.edu.cn

Abstract: A multilevel contact oxidation system was applied in a pilot-scale experiment to treat the automobile painting wastewater, which had poor biodegradability and contained high concentration of Chemical Oxygen Demand (COD). The wastewater used for this experiment study was the actual painting wastewater which had been pre-treated by the physic-chemical process, and its Biological Oxygen Demand (BOD_5) /COD was less than 0.1,COD concentration was 800~1500mg/L. The results showed that the multilevel contact oxidation system could efficiently degrade the COD of the painting wastewater. When the experimental system kept stable operation, the total removal rate of COD and suspended solid (SS) were 84% and 82.5% respectively with the Hydraulic Retention Time (HRT) of 8 hours. Meanwhile, this system had a strong ability to resist the impact of COD concentration change. The COD concentration of final treated wastewater was less than 500 mg/L, which could reach the factory discharge requirement for the paint shop. Besides, this system with simple structure was able to reduce the excess sludge production greatly, which would reduce much cost for the treatment of painting wastewater.

1. Introduction

Before an automobile being painted in auto manufacturers, several technological procedures are needed. Such as, degreasing, activation, passivation, electrophoresis, acid cleaning, alkaline cleaning, etc. During these processes, much wastewater will be generated, which contains various and complicated organic substance, like surface active agent、electrophoretic paint and fossil oil, and also contains a certain amount of heavy metal [1-3]. Except the inorganic matter, the complicated organic substance is very hard to biodegrade. B/C stands for the biodegradability indicator of paint shop wastewater and it is less than 0.1 in general [4].

At present, the physicochemical and biochemical technology has become the main treatment process for domestic automobile painting wastewater. And classification pretreatment-hydrolysis acidification-aerobic treatment process is relatively common used in automobile painting wastewater treatment. The common treatment methods for domestic automobile painting wastewater are hydrolysis acidification-SBR (such as: Iveco car factory); hydrolysis acidification-biological aerated filter (such as: Jiangling

Group); hydrolysis acidification-single stage contact oxidation (such as: Changchun car factory); hydrolysis acidification-MBR (such as: Changchun car accessories factory).

The specific treatment process should be chose according to some factors, such as, the painting wastewater characteristics, discharge conditions, cover space, outlet requirements, etc.

In this paper, the car factory requires that the fossil oil、heavy metal and COD must be reduced and reach the internal targets before discharging this kind of painting wastewater into the general sewage treatment station inside the factory, which is aimed to reduce the influence on the subsequent wastewater treatment and outlet index in the general sewage station. All kinds of wastewater in paint shop are treated by a series of physic-chemical methods and the COD concentration of the mixed wastewater at the end of the system is 800~1500mg/L. However, the COD concentration discharge standard in paint shop is performed according to the level 3 standard of maximum allowable discharge concentration of the second type of pollutant in the integrated wastewater discharge standard《DB8978-1996》, namely, COD should be less than 500mg/L.

For the above problem, comprehensively considering the treatment effect, cover space and running cost in practical engineering, a multilevel biological contact oxidation system is applied to retreat the wastewater which has been pre-treated by the physic-chemical process in paint shop. Figure 1 is the flowchart of the original physic-chemical process with new added biological process.

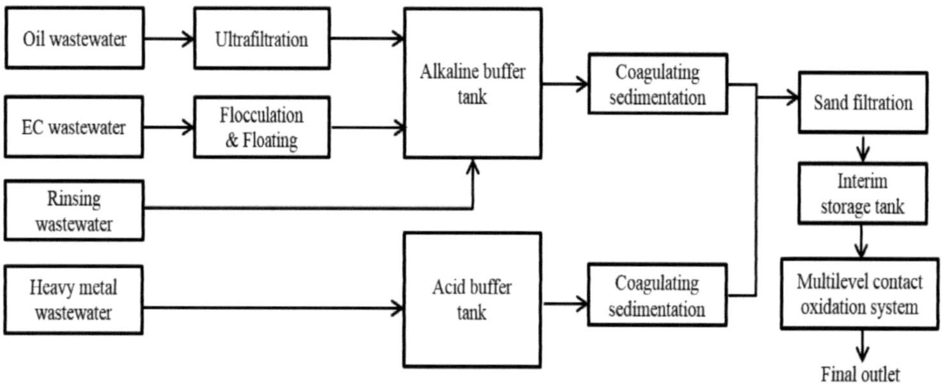

Figure 1. Flowchart of the painting wastewater treatment process.

2. Experiment wastewater

The water used for this experiment is from that interim storage tank, as shown in figure 1. Because of the process particularity in paint shop, the change of wastewater quality is not stable. So the COD concentration of wastewater used for the experiment is also changing randomly. Besides, the biodegradability of painting wastewater is poor and B/C is only about 0.1. The situation of the painting wastewater quality is shown in table 1.

Table 1. Quality of wastewater used for experiment

Main pollutant	PH	COD (mg/L)	BOD (mg/L)	TP (mg/L)	TN (mg/L)	NH_3-N (mg/L)	SS (mg/L)
Test value	6.5~ 8.5	800~ 1500	50~100	0.3~0.9	0~20	1~10	50~200

3. Biological experiment device

The experiment system mainly includes inlet tank, multilevel contact oxidation reactor and outlet tank, as shown in figure 2.

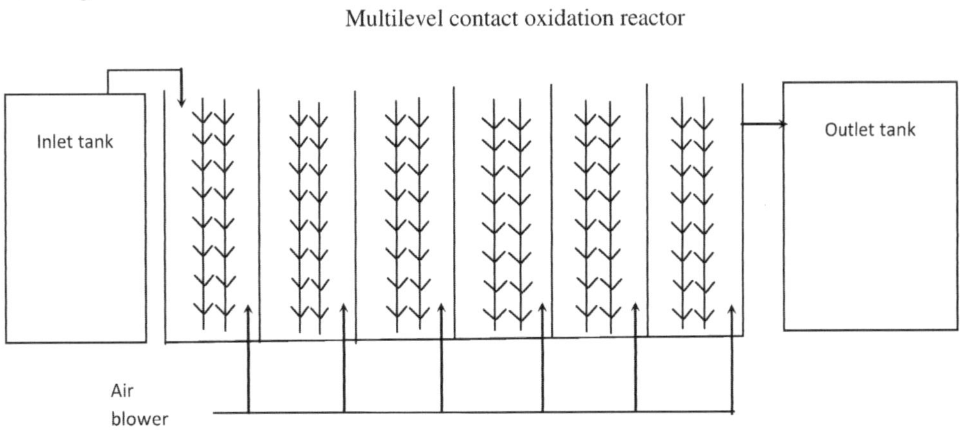

Figure 2. Experiment system of multilevel contact oxidation.

The multilevel contact oxidation reactor includes six biological tanks and its total effective volume for the experiment is 80L. The stuffing frame is placed in each biological tank and the special stuffing which can be used to cultivate microbial membrane efficiently is placed on the stuffing frame. The aeration system with different intensities is placed on the bottom of each biological tank, which can bring air into the biochemical tank through the air blower, providing enough oxygen to the microbial. Besides, the aeration intensity and dissolved oxygen concentration can be adjusted according to the changes of water quality and the growth state of biofilm [5].

4. Experimental start-up and process

The inoculation sludge used for the experiment is the sludge from the secondary sedimentation tank of the general sewage treatment plant in the car factory. The MLSS (Mixed Liquid Suspended Solids) of the initial dosing sludge is about 3000 mg/L. 25L active sludge, 30L painting wastewater and 25L clean water is evenly mixed and then it's fed to the biological tank where is placed with special biological stuffing. Two days later, the whole surface of biological stuffing in each tank is covered evenly by active sludge and a layer of mud-yellow biofilm is growing, which can prove that the biofilm formation is successful.

Later, the experiment water begins to flow into the biological tank constantly from the inlet tank through the water pump. The flowrate starts with 80L/d and gradually increases to 240L/d, which is controlled by the flowmeter. That's to say, the HRT starts with 24 hours and gradually reduces to 8 hours. It has taken about 30 days totally during the experiment commissioning and running.

The biofilm in each biological tank is keeping growing and getting thicker over time until reaching a stable state. Besides, the biofilm in the first biological tank is the thickest and the biofilm texture is the closest. The biofilm thickness and texture in the later biological tanks are decreasing in turn. The appearance color of stuffing in each biological tank is also changing constantly, as shown in figure 3. The appearance color of stuffing in the start stage is mud-yellow at the beginning, then slowly turns into deep tan, and finally becomes black. The appearance color of stuffing in the middle stage is mud-yellow at the beginning, then slowly turns into tan. The appearance color of stuffing in the final stage is mud-yellow.

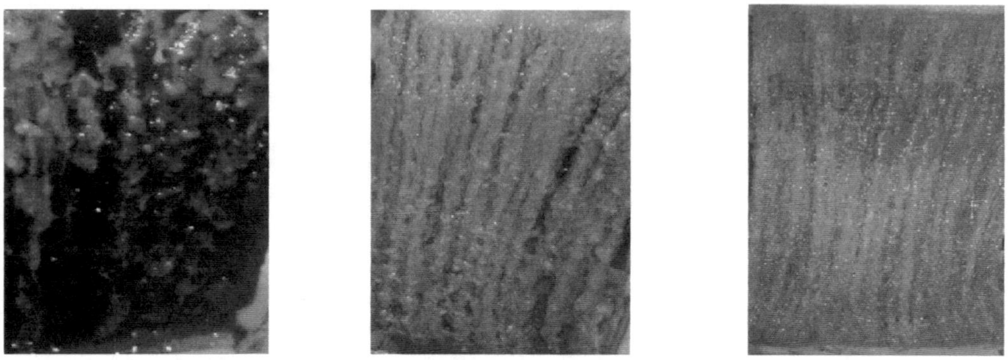

(a) The start stage (b) The middle stage (c) The final stage

Figure 3. The appearance situation of stuffing in the biological system.

5. Experiment results and analysis

5.1 Organic matter removal

During the experiment, every day take water sample regularly and test the COD value. Take samples at three special points during the whole biological system: the start stage, the middle stage and the last stage.

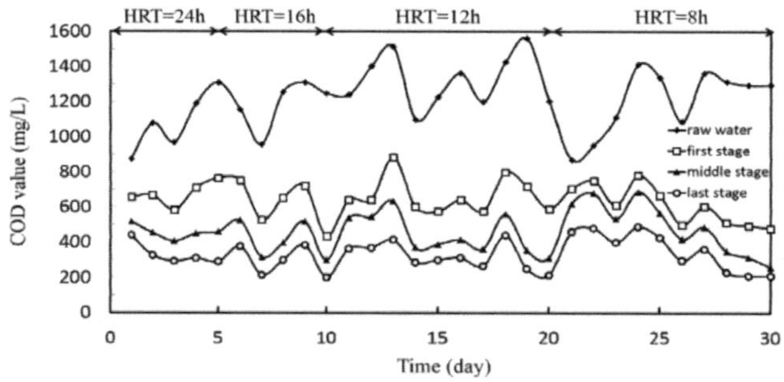

Figure 4. Variation of COD concentration on condition of different HRT.

As shown in figure 4, under the conditions of different COD loads and different HRT, the COD concentration in the last stage is less than 500mg/L and can reach the discharge standard stably.

It can also be seen that an obvious COD concentration difference has formed between every biological tank. This is because that the wastewater flows through each biological tank in turn, the difference of pollutant composition and concentration in each biological tank will result in a different kind of microbial generating accordingly and forming a biologic chain. The microbial with different species and structures in each tank can efficiently decompose the refractory organic matters step by step, and finally the COD concentration is able to reach the discharge standard stably [6-7].

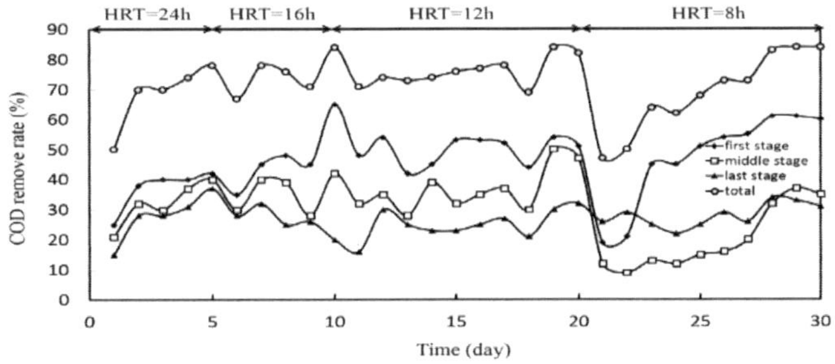

Figure 5. Variation of COD remove rate on condition of different HRT

As shown in figure 5, under the condition of different HRT, the COD remove rate in each stage of the multilevel contact oxidation system decreases step by step. Besides, the change of inlet flowrate has the greatest influence on the COD remove rate in the first stage and the least influence on the COD remove rate in the last stage.

This is because that the microbial in the first biological tank plays a main role in degrading the COD. Since the COD concentration of wastewater entering the first biological tank is the highest, the microbial in the first biological tank can grows fastest and degrades the COD most quickly under the suitable dissolved oxygen (2.0-4.0mg/L) and water temperature (25-30°C). When the inlet flowrate is changed, the wastewater with high COD concentration will first enter the first biological tank. So the microbial in the first biological tank will be most greatly affected by the concentration impact and the degradation efficiency will also appear a larger fluctuation accordingly [8].

5.2 Impact resistant properties

During the experiment, the COD concentration range of influent is 800~1500mg/L. When the HRT is set as 24h, 16h, 12h and 8h successively, the experimental test results show that the COD concentration of the final effluent can still stay in 300~500 mg/L stably.

When the HRT is shortened every time, the COD removal rate in each stage decreases first, then increases, finally tends to be stable. That's because when the HRT is shortened, the inlet flowrate will be increased, the microbial in each biological tank needs time to adapt to the changed environmental conditions. The experiment results can show that this multilevel contact oxidation system can adjust to the changed HRT in shorter time and quickly achieve the stable operation. After the biological system achieving stable operation, the total COD remove rate can reach 84% with HRT of 8 hours.

5.3 Excess sludge production

This experiment has been running for 30 days and treated 5000L painting wastewater, but the excess sludge production is only less than 0.5kg.

When the biological system reach stable operation with the HRT of 8 hours, the suspended solid (SS) of influent and effluent has been tested, as shown in figure 6. The average remove rate of SS is about 82.8%.

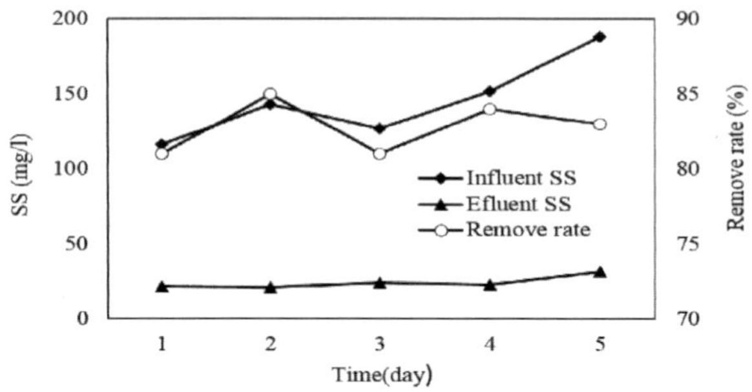

Figure 6. Variation of SS concentration and SS removal rate with HRT of 8 hours

The reasons why this multilevel biological contact oxidation system can efficiently reduce the excess sludge production are as follows:

Firstly, the difference of pollutant composition and concentration in each biological tank will result in a different kind of microbial generating accordingly, and the microbial in each biological tank is changing and evolving continually, thereby forming a food chain. Higher trophic level microbial can prey on the low trophic level microbial, such as the bacteria and the dead microbial, thus decreasing the sludge yield.

Secondly, energy was lost at every transition in a natural food chain because part of the organic material is converted into carbon dioxide and water [9]. So, the yield coefficient of microbial will be reduced. Accordingly, the sludge yield will also be reduced.

Thirdly, Higher microbial is easy to survive and can keep endogenous metabolism [10], which can decrease the yield of dead microbial, namely, decrease the sludge yield.

6. Conclusions

The experiment results show that the multilevel contact oxidation system is able to reduce the COD concentration of painting wastewater by cultivating activated sludge and controlling process conditions. During the whole experiment commissioning and running, the painting wastewater after pretreatment by physicochemical method is retreated by the multilevel biological contact oxidation system, and the COD concentration of outlet water can reach the internal discharge standard.

When the HRT is shorten every time, the system can quickly achieve stable operation again, so the system has strong impact resistance. Besides, the system can effectively reduce the excess sludge production.

Comparing the traditional biological methods for treating automobile painting wastewater, the multilevel contact oxidation system has obvious advantages. Firstly, it can efficiently reduce the excess sludge production and greatly reduce the sludge treatment cost. Secondly, the reactor structure is simple, the cover space is small, the operation is easy and the running is stable. Thirdly, it has high treatment efficiency and can decompose the refractory organic matters in painting wastewater.

This multilevel contact oxidation method can greatly reduce the treatment cost of automobile painting wastewater, and it has good economic and social benefits. The experiment results have important reference value for the automobile painting wastewater treatment.

References

[1] Tian Y P, Kong S S. Study on automobile painting-wastewater treatment with the process technology of pretreatment-air flotation-SBR[J]. Advanced Materials Research, 2012,550-553:2237-2240.

[2] Huang C, Liu Y Y, Luo Y. Research status of dyeing wastewater treatment [J]. Chongqing University, 2001, 24 (6), 139-142.

[3] Chen C L, Zhang J A, Li X P,et al. Treatment of wastewater coming from painting processes:application of conventional and advanced oxidation technologies[J]. Ozone Science & Engineering, 2005, 27(27):279-286

[4] Shaw C B , Carli ell C M , Wheat ley A D. Anaerobic/aerobic treatment of colored textile effluents using sequencing batch reactors[J]. Water Research, 2002, 36(8):1993-2001.

[5] Li X D, Wang Q H, Li Q F. Study on treatment of simulant dairy processing wastewater using FCR multilevel contact oxidation system[J]. Environmental science, 2007, 28(9):2020-2024.

[6] Du T X, Fei Q Z, Xu Z. Study of soybean wastewater treatment with eight-series biological contact oxidation system[J]. Journal of Dalian Jiao tong university, 2008, 29(3):67-69.

[7] Qiu G L, Cheng J G, Xiang L C. Pilot-Scale Test of Biological Contact Oxidation Process for Decentralized Wastewater Treatment [J], China Water &Wastewater, 2007, 23(5):78-81.

[8] Rocher M, Goma G, Begue A P.Towards a reduction in excess sludge production in activated sludge process: biomass physicochemical treatment and biodegradation [J]. Appl Microbio Biotechnol, 1999, 6(51):883-890.

[9] Ghyoot W, Verstraete W. Reduced sludge production in a two-stage membrane-assisted bioreactor [J].Wat Res, 1999, 34(1):205-215.

[10] Divyalakshmi P, Murugan D, Sivarajan M. In situ disruption approach on aerobic sludge biomass for excess sludge reduction in tannery effluent treatment plant [J]. Chemical Engineering Journal, 2015, 276:130-136.

Adsorption Characteristics of Macroporous Resin for Oil Removal from Desulphurization Wastewater on Board

Tie Li, Chen Chen, Qi Jin, Jiao Zhao, Xiaojia Tang and Yimin Zhu

Collaborative Innovation Center for Vessel Pollution Monitoring and Control, Dalian Maritime University, Dalian, Liaoning 116026, China

ntp@dlmu.edu.cn

Abstract. According to our previous results on the magnesium-based exhaust gas cleaning system (Mg-EGCS), PAHs and total oil content were the main factors affecting the COD in the wastewater. In this work, three kinds of adsorption materials were investigated and macroporous resin was selected for oil removal. The effects of the dosage of macroporous resin, adsorption time and the flow rate were studied, and thermodynamics equation was used to characterize the adsorption process. The results showed that macroporous resin is a good candidate for oil removal from desulphurization wastewater on board, and the COD after treatment can meet the discharge criteria set by the International Maritime Organization (IMO).

1. Introduction

A novel magnesium-based exhaust gas cleaning system (Mg-EGCS), which adopts magnesium oxide or magnesium hydroxide as absorbent, has been reported in our previous studies, and after a long time operation on board, Mg-EGCS has been proven to meet the IMO sulfur emission requirements [1]. However, due to the heavy fuel oil used in the diesel engine, the exhaust gas also contains unneglectable products from incomplete combustion, such as aromatic and naphthenic compounds, which would be washed into the wastewater in the desulphurization process. If such wastewater was discharged without treatment, it would cause serious seawater pollution.

The composition of the wastewater from the Mg-EGCS has been analyzed in our laboratory [2], and the results showed that the COD in the wastewater was beyond the limitation based on the washwater discharge criteria. The PAHs and total oil content were the main factors affecting the COD in the wastewater.

Various methods of oil removal have been developed. Among them, physical method was simple but only suitable for the wastewater which contained high oil content [3], while chemical method and biology method showed better oil removal efficiency, but the processes were usually complicated [4-6].

Adsorption method was commonly used for oily wastewater treatment [7]. The typical adsorption materials include activated carbon, fly ash, bentonite, cotton fibers adsorbent, ceramsite filter and adsorption resin, etc.

According to the characteristics of the desulphurization wastewater and the special requirements of the ship wastewater treatment, in this work, three kinds of adsorption materials were investigated and macroporous resin was selected for oil removal. The effects of the dosage of macroporous resin, stirrer speed and the flow rate were studied, and thermodynamics equation was used to characterize the adsorption process.

IOP Publishing
doi:10.1088/1755-1315/51/1/012019

2. Materials and method

2.1. Preparation of simulated oil wastewater

The simulated oily wastewater was prepared as follows: certain amount of fuel oil was put into a beaker, and mixed with sea water; after stirring for 24 hours, the mixture was kept stand for 5 hours. It should be diluted to about 140 mg/L if the oil content in water is too high.

2.2 Properties of macroporous resin

The physicochemical properties of macroporous resin are shown in Table 1.

Table 1. The physicochemical properties of macroporous resins

Item	Parameters
Wàter content (%)	55~65
Particle size (μm)	250~600
Specific surface area (m^2/g)	590
Pore volume (mL/g)	1.3
Relative density	1.01
Apparent density (g/L)	680

2.3 Analysis methods

The COD was analyzed according to the standard methods (APHA, AWWA and WEF, 1999). The PAHs and total oil were determined by liquid chromatography (Agilent 1260) and infrared spectrometer (JDS-100, Jilin North Light Technology Co., Ltd., China), respectively. The oil removal rate D_0 was calculated as follows:

$$D_0 = \frac{\rho_0 - c_e}{\rho_o} \times 100\% \tag{1}$$

where ρ_0 is the initial oil concentration in water sample (mg/L), C_e is the equilibrium oil concentration (mg/L).

3. Results and discussion

3.1 Comparison of adsorption performance

Cotton fibers adsorbent 0.5 g, ceramist filter 3 g and macroporous resin 1.0 g were stirred with 50 mL of desulfurization wastewater with initial oil concentration of 18.638 mg/L in a beaker, respectively. After 3 hours, the residual oil content in the solution was measured. Such parallel experiments were repeated for 3 times.

The effects of oil removal of various adsorbent materials on desulfurization wastewater were showed in Table 2. The oil removal rate of cotton fibers adsorbent and ceramist filter was 8.62% and 4.05%, respectively, which was less than that of macroporous resin with the oil removal rate of 51.22%.

Table 2. The adsorption performance of various adsorbent materials on desulfurization wastewater

Adsorbent material	cotton fibers adsorbent	ceramist filter	macroporous resin
Residual oil content	17.045	17.826	9.295
C_e(mg/L)	16.832	18.295	8.931
	17.221	17.532	9.051
Average C_e(mg/L)	17.032	17.884	9.092
Removal rate (%)	8.62	4.05	51.22

3.2 The adsorption experiments of macroporous resin

3.2.1 The effect of the dosage of macroporous resin. In order to determine the effect of oil removal, the dosage of macroporous resin was kept as 4.0, 5.0, 6.0, 7.0, 8.0, 12.0, 16.0, 24.0, 28.0, 36.0, 44.0 and 60.0 g/L, respectively, and the experimental conditions were kept as the initial oil concentration of 140 mg/L and the stirrer time of 4 hours.

As shown in Figure 1, the oil removal rate increased with the increase of the macroporous resin dosage, and the removal rate reached 94% when the dosage was increased to 44.0 g/L. As the dosage of macroporous resin raised from 4.0 to 8.0 g/L, a significant increase of the removal rate could be observed. While in the range of 12.0 g/L to 44.0 g/L, there was only a slight increase of the removal rate.

When the dosage was in the range of 4.0 ~ 8.0 g/L, the macroporous resin can be fully contacted with the water by stirring, therefore, the oil removal rate increased rapidly. When the dosage increased from 12.0 g/L to 44.0 g/L, the density of macroporous resin decreased, which cannot fully contact with water.

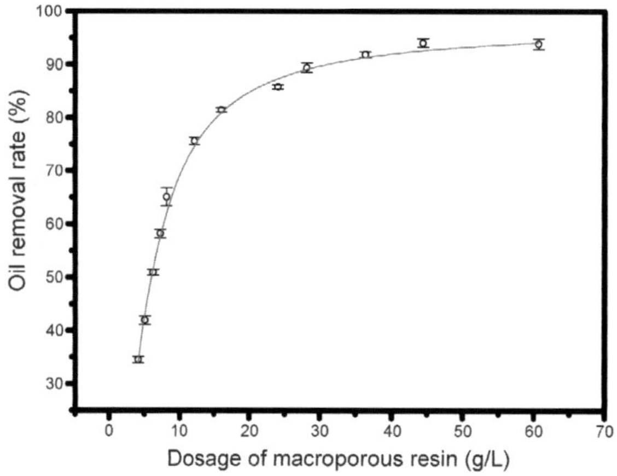

Figure 1. The effect of the dosage of macroporous resin.

3.2.2 The effect of adsorption time. Figure 2 showed the effect of adsorption time on the oil removal rate, when it was varied from 60 to 240 min. The other parameters were kept constant for all trials with the initial oil concentration of 140 mg/L, dosage of macroporous resin of 7.0 g/L and a stirring speed of 1,100 rpm. The maximum oil removal rate reached to 58% with the adsorption time of 165 min, then only a slight increase of removal rate was observed with further increase of the adsorption time.

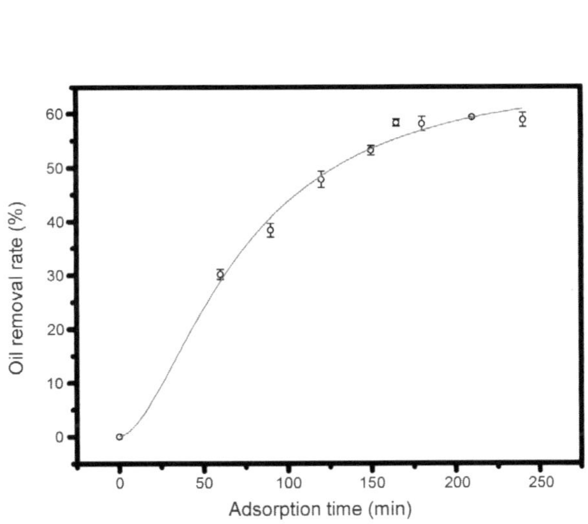

Figure 2. The effect of the adsorption time.

3.2.3 The effect of flow rate. The macroporous resin was packed in a chromatographic column with an inner diameter of 20 mm and a height of 400 mm. The flow rate was adjusted to 4.5 mL/min, 7.5 mL/min and 10 mL/min, respectively. Then the oil content was measured every ten minutes. In each batch of experiment, the macroporous resin and the initial oil concentration were kept as 10 g and 140 mg/L, respectively. The effect of flow rate on the oil removal rate was shown in Figure 3.

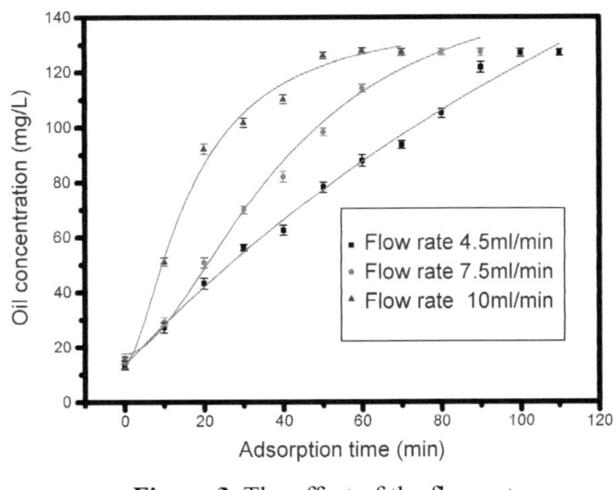

Figure 3. The effect of the flow rate

It can be seen from Fig. 3 that the permeation time decreased with the increase of the flow rate from 4.5 mL/min to 10 mL/min. When the flow rate was 4.5mL/min, 7.5 mL/min and 10 mL/min, the penetration time was 90 min, 70 min and 50 min respectively. The decrease of the flow rate will increase the contact time with the macroporous resin, and be helpful for the absorption of oil.

3.2.4 Adsorption isotherm. Various isotherm equations have been used to describe the equilibrium nature of adsorption, including the Freundlich and Langmuir equations.

The Langmuir isotherm which is applicable for monomolecular layer adsorption, has been extensively used for characterizing the adsorption of heavy metals, dyes, and organic pollutants [8]. This isotherm is described as homogeneous adsorption, assuming that all the adsorption sites have equal adsorbate affinity and that the adsorption at one site does not affect the adsorption at an adjacent site [9-11]. The Langmuir isotherm is used to obtain a maximum adsorption capacity produced from the complete monolayer coverage of the adsorbent surface. The isotherm equation is represented as:

$$q_e = \frac{bQ_m C_e}{1+bC_e} \tag{2}$$

where Q_m and b are coefficients, q_e is the weight adsorbed per unit weight of adsorbent, and C_e is the oil concentration in the water at equilibrium. The value of Q_m and b were obtained from slope and intercept of linear curve presented in Figure 4.

The Freundlich isotherm is given by the following equation:

$$q_e = KC_e^{1/n} \tag{3}$$

By taking logarithms of both sides of equation (3) and rearranging, equation (4) was obtained.

$$\log q_e = \log K - n \log C_e \tag{4}$$

where n is the Freundlich constant, K is the adsorption coefficient, q_e is the weight adsorbed per unit weight of adsorbent, and C_e is the equilibrium oil concentration in the water. The values of K and $1/n$ were calculated from the intercept and slope of linear curve presented in Figure 5.

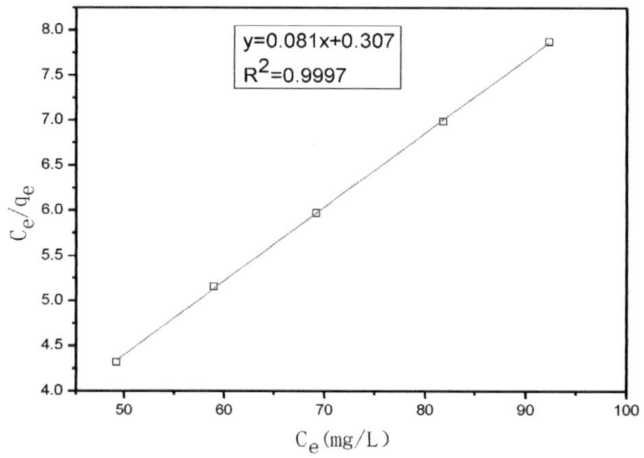

Figure 4. Langmuir models for oil removal rate.

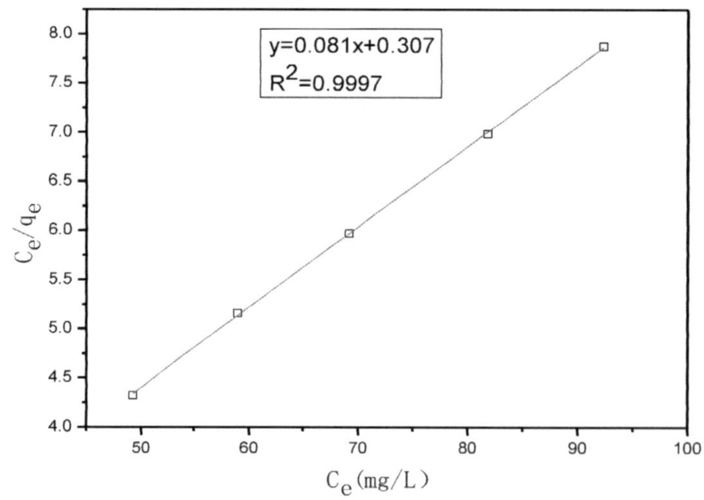

Figure 5. Freundlich models for oil removal rate.

The constants related to the aforementioned isotherm models were calculated (Table 3), and the results indicated that the data obtained throughout the experiments were well fitted to Langmuir isotherm model, with a correlation coefficient of 0.99. Moreover, the adsorption capacity of macroporous resin for the removal of oil was determined to be 11.727 mg/g; which represents an improved capacity compared with the capacities of other well-known adsorbents.

Table 3. Isotherm constants for oil adsorption

Langmuir			Freundlich		
Q_m	b	R^2	K	n	R^2
0.264	12.345	0.999	8.561	-0.051	0.933

4. Conclusion

The present study shows that macroporous resin is an effective adsorbent for the removal of oil from desulphurization wastewater. The adsorption capacity of the macroporous resin was 12.35 mg/g and the equilibrium adsorption time was about 3 h. The results of dynamic adsorption experiments showed that with the increase of flow rate, the adsorption time was reduced and the penetration time was shortened. Isotherm model studies indicated that the Langmuir isotherm model fits the equilibrium data very well for the macroporous resin adsorption system. In conclusion, the investigations indicated that macroporous resin is effective for oil adsorption due to its special adsorption characteristics. The results of this study are useful for the design of treatment process for the removal of oil in the desulphurization wastewater.

Reference

[1] Tang, X.J., Li, T., Yu, H. and Zhu, Y.M. 2014 *Ocean Eng.* **76** 98
[2] Li Tie, Yu Hang, Li Ye-ye, Liu Quan, Chen Chen, Guo Lin and Zhu Yi-min 2015 Int J *Environ. & Poll.* **53** 1
[3] Xiao-ru Cheng and Ying-wen xue 2002 *Environ. Sci. & Tech.* **3** 38
[4] Oguz E, Keskinler B and Celik Z 2005 *Dyes Pigments* **64** 01
[5] Zheng-qing Yang, Li-Xia Yang, Yi-ning Hou and An Wang 2006 *Environ. Sci. & Man.* **31** 119
[6] Yang Bai 2003 *Sci-Tech Information Development & Economy* **13** 113
[7] Chang-ying Guo, Peng-cheng Zhao and Jing Xiao 2010 *Environ. Sci. & Man.* **3** 96
[8] M. Ghaedi, S. Heidarpour, S.N. Kokhdan, R. Sahraie, A. Daneshfar and B. Brazesh 2012 *Powder Technol.* **228** 18.
[9] S. Chatterjee, S. Chatterjee, B.P. Chatterjee and A.K. Guha 2007 *Colloids Surfaces A* **299** 146
[10] D. Mohan, K.P. Singh, G. Singh and K. Kumar 2002 Indus. Eng. Chem. Res. **41** 3688
[11] Y.E. Benkli, M.F. Can, M. Turan and M.S. Celik 2005 Water Res. **39** 487.

Freshwater Sediment Characterization Factors of Copper Oxide Nanoparticles

Yubing Pu[1,2], Bertrand Laratte[2,3,4] and Rodica Elena Ionescu[1]

[1]Laboratoire de Nanotechnologie et d'Instrumentation Optique, ICD, UMR CNRS 6281, Université de Technologie de Troyes, 10004 Troyes, France
[2]Centre de Recherches et d'Etudes Interdisciplinaires sur le Développement Durable, ICD, UMR CNRS 6281, Université de Technologie de Troyes, 10004 Troyes, France
[3]Institute of Mechanics and Engineering, I2M, UMR-CNRS 5295, Arts et Métiers ParisTech, 33405 Talence, France
[4]APESA, Technopole Izarbel, 64210 Bidart, France

E-mail: yubing.pu@utt.fr

Abstract. Wide use of engineered nanoparticles (ENPs) is likely to result in the eventually accumulation of ENPs in sediment. The benthic organisms living in sediments may suffer relatively high toxic effects of ENPs. This study has selected copper oxide nanoparticles (nano-CuO) as a research object. To consider the impacts of spatial heterogeneity on ENPs toxicity, the characterization factor (*CF*) derived from life cycle assessment (LCA) methodology is used as an indicator in this study. A nano-specific fate model has been used to calculate the freshwater sediment fate factor (*FF*) of nano-CuO. A literature survey of the nano-CuO toxicology values has been performed to calculate the effect factor (*EF*). Seventeen freshwater sediment CFs of nano-CuO are proposed as recommended values for subcontinental regions. The region most likely to be affected by nano-CuO is northern Australia (*CF* of $21.01 \cdot 10^3$ CTU_e, comparative toxic units) and the least likely is northern Europe and northern Canada (*CF* of $8.55 \cdot 10^3$ CTU_e). These sediment CFs for nano-CuO could be used in the future when evaluating the ecosystem impacts of products containing nano-CuO by LCA method.

1. Introduction

The volume of production of nanomaterials (NMs) have grown extensively over the last decade.[1] NMs are widely used in many household products including cosmetics, sunscreens, textiles etc.[2] Although people benefit a lot in using NMs, the potential negative impacts of NMs are also raised concerns. Many researches demonstrated the toxicity of NMs to environment.[3, 4] As a principal member of NMs, the engineered nanoparticles (ENPs) are widely used and likely to be released into ecosystem.[1] It was reported that the ENPs are usually released into the freshwater through the wastewater treatment plant.[5] Freshwater sediment is one of the main fate of suspended ENPs in freshwater.[6] Thus, the benthic organisms living in sediments may suffer relatively high toxic effects of ENPs.[7]

Generally, the bio-experiment is one effective way to investigate the toxicity of ENPs on organisms. However, most experiments have difficulty in assessing the impacts of spatial heterogeneity on ENPs toxicity. A recently published article indicated that the hydrogeology has significant influences on the impacts of ENPs because of the different residence time of ENPs in different freshwaters.[6] In order to assess the impacts of hydrogeology on nanotoxicity to sediment, characterization factor (*CF*) is used as an indicator in this study. The *CF* is derived from life cycle assessment (LCA) and reflects the relative importance of each component in life-cycle inventories.[8] It can be calculated under the framework of a scientific consensus environment model, USEtox model, which is recommended by many international organizations (e.g. European Commission, etc.).[9]

This study has selected copper oxide nanoparticles (nano-CuO) as a research object due to their relatively high toxicity than other metal oxide nanoparticles (e.g. nano-ZnO, nano-TiO_2, etc.)[10] The aim of this study is to propose the freshwater sediment CFs of nano-CuO for different regions. A nano-specific fate model with 17 regional-specific input parameters has been applied to calculate the sediment fate factors (FFs) of nano-CuO.[6] In addtion, a literature survey of nano-CuO toxicology values has been performed to calculate the effect factor (*EF*). Finally, seventeen freshwater sediment CFs of nano-CuO are proposed as recommended values for subcontinental regions.

2. Methods

2.1. General framework

The USEtox model proposed a method to calculate the ecotoxicological characterization factor (*CF*) of a substance for freshwater ecosystems. In this study, to calculate the *CF* for freshwater sediment, a small modification of the method is needed. The research object should be changed to the sediment compartment. The adapted characterization factor (*CF*, unit: PAF·m^3·day·kg^{-1}) of a substance for freshwater sediment is the product of three factors: fate factor (*FF*, unit: day), exposure factor (*XF*, unit: dimensionless) and effect factor (*EF*, unit: PAF·m^3·kg^{-1}):[9]

$$CF = FF \cdot XF \cdot EF \tag{1}$$

The *FF* represents the residence time of the substance in sediment and the *XF* is the bioavailability of a substance. The *EF* reflects the change in the potentially affected fraction (PAF) of species because of the change in substance concentration in sediment. In practice, the unit of *CF* is defined as comparative toxic units (CTU_e) for ecosystem.

2.2. Calculation of fate factor

2.2.1. *Fate model concept* A nano-specific fate model with 17 regional-specific input parameters has been used to calculate the sediment fate factors (FFs) of nano-CuO.[6] Figure 1 shows the simplified fate model of nano-CuO in freshwater-sediment ecosystem. It was assumed that there is no direct ENPs emissions to sediment.[6] The following behavior as shown in figure 1 is considered as the removal processes for sediment: (1) resuspension from sediment to water; (2) burial into the deep sediment; (3) horizontal bed load transfer to other sediment compartment.

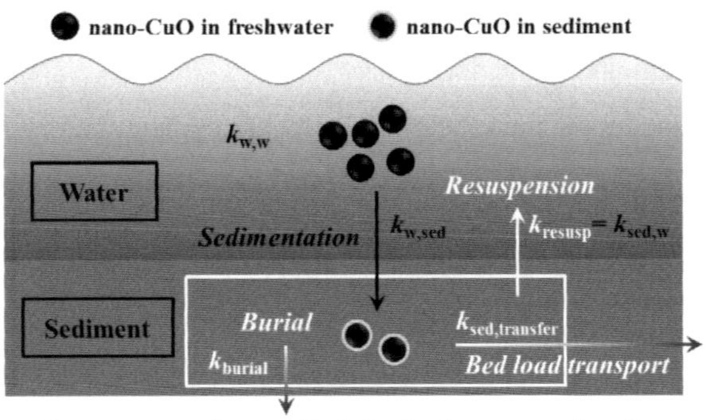

Figure 1. The behavior and transport of nano-CuO in freshwater and sediment with the corresponding rate constants accounted (Modified from Pu, et al.).[6]

2.2.2. *Sediment fate factor* The residence time of nano-CuO in freshwater sediment ($FF_{sed,sed}$, day) is calculated as equation 2:

$$FF_{sed,sed} = \frac{k_{w,w}}{k_{w,w} \cdot k_{sed,sed} - k_{w,sed} \cdot k_{sed,w}} \tag{2}$$

The total removal rate of nano-CuO in sediment ($k_{sed,sed}$, s^{-1}) is expressed as a sum of the rate constant for horizontal bed load transfer ($k_{sed,transfer}$, s^{-1}), burial (k_{burial}, s^{-1}) and resuspension (k_{resusp}, s^{-1}):

$$k_{sed,sed} = k_{sed,transfer} + k_{burial} + k_{resusp} \tag{3}$$

The detailed calculation methods and the input values for the above elements could be found in the study of Pu, et al.[6] The nano-CuO density of 6320 $kg \cdot m^{-3}$ was used in the calculation. If not specified, the *FF* refers to $FF_{sed,sed}$ in this study.

2.3. Calculation of exposure factor

In USEtox, the freshwater exposure factor (*XF*) is defined as the fraction of a chemical dissolved in freshwater.[9] In this study, it is assumed that the CuO nanoparticles are all bioavailable and dispersed homogeneous in the freshwater sediment. Thus, here *XF* is set to be 1 to avoid significant errors in nano-CuO *CF* predictions.

2.4. Calculation of effect factor

The aquatic ecotoxicological effect factor (*EF*) is calculated based on geometric means of single species EC50 value in USEtox.[9] EC50 ($kg \cdot m^{-3}$) is the effective concentration of a substance at which 50% of a population (e.g. fish) displays a phenomenon (e.g. mortality). In this study, due to the research object is sediment, the $EC50_{ref}$ from reference has a unit of kg nano-CuO per kg sediment. Therefore, a unit conversion is necessary to obtain the EC50 in unit of kg nano-CuO per m^3 sediment. In addition, when chronic EC50 value is unavailable, an acute to chronic ratio (*ACR*, 15 in this study) will be used to derive the chronic-equivalent EC50.

$$EC50_{chronic} = \frac{EC50_{ref} \cdot \rho_{sed}}{ACR} \tag{4}$$

where ρ_{sed} (kg sediment per m^3 sediment) is the bulk density of freshwater sediment. Here, the value of 1230 $kg \cdot m^{-3}$ has been applied.[6, 9] The obtained $EC50_{chronic}$ was used to calculate the sediment *EF* of nano-CuO. The detailed calculation methods of *EF* could be found in the study of Pu, et al.[6]

3. Results and discussions

3.1. Fate factors of nano-CuO

As shown in figure 2, a low persistence of nano-CuO in the sediment (~1218 day) occurs in northern Europe and northern Canada regions (W12), while high *FF* values (~2991 day) is observable in northern Australia (W3). In the following calculation of CFs, these values of FFs were used.

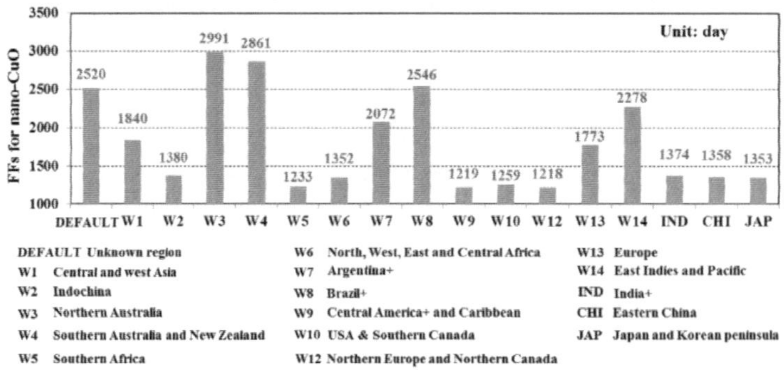

Figure 2. The recommended values of sediment FFs for nano-CuO in 17 subcontinental regions. (Further details of the countries within each region can be found in the study of Shaked.[11])

The transport rate constants of nano-CuO between freshwater and sediment for 17 subcontinental regions are listed in Table 1 to explain the differences of *FFs*. For example, the transport rate of nano-CuO from freshwater to sediment ($k_{w,sed}$) for W3 region is more than 5 times faster than that for W12 region ($2.07 \cdot 10^{-5}$ versus $3.77 \cdot 10^{-6}$ s^{-1}), while the inverse transport rate ($k_{sed,w}$) for W3 region is less than one third of that for W12 region ($3.32 \cdot 10^{-9}$ versus $1.00 \cdot 10^{-8}$ s^{-1}). It's the apparent reason of longer residence time in sediment of W12 region than W3 region. The underlying cause is the the subcontinental difference of freshwater-sediment parameters.

Table 1. The transport rate constants of nano-CuO between freshwater and sediment for 17 subcontinental regions (Unit: s^{-1}).

	$k_{w,w}$	$k_{w,sed}$	$k_{sed,sed}$	$k_{sed,w}$		$k_{sed,w}$	$k_{sed,w}$	$k_{sed,w}$	$k_{sed,w}$
DEF	$2.57 \cdot 10^{-5}$	$2.07 \cdot 10^{-5}$	$1.17 \cdot 10^{-8}$	$8.87 \cdot 10^{-9}$	**W9**	$8.14 \cdot 10^{-6}$	$3.11 \cdot 10^{-6}$	$1.33 \cdot 10^{-8}$	$9.97 \cdot 10^{-9}$
W1	$9.99 \cdot 10^{-6}$	$4.97 \cdot 10^{-6}$	$9.43 \cdot 10^{-9}$	$6.30 \cdot 10^{-9}$	**W10**	$8.12 \cdot 10^{-6}$	$3.11 \cdot 10^{-6}$	$1.28 \cdot 10^{-8}$	$9.33 \cdot 10^{-9}$
W2	$9.83 \cdot 10^{-6}$	$4.78 \cdot 10^{-6}$	$1.32 \cdot 10^{-8}$	$9.87 \cdot 10^{-9}$	**W12**	$8.78 \cdot 10^{-6}$	$3.77 \cdot 10^{-6}$	$1.38 \cdot 10^{-8}$	$1.00 \cdot 10^{-8}$
W3	$2.60 \cdot 10^{-5}$	$2.07 \cdot 10^{-5}$	$6.51 \cdot 10^{-9}$	$3.32 \cdot 10^{-9}$	**W13**	$9.05 \cdot 10^{-6}$	$4.04 \cdot 10^{-6}$	$9.20 \cdot 10^{-9}$	$6.00 \cdot 10^{-9}$
W4	$2.58 \cdot 10^{-5}$	$2.07 \cdot 10^{-5}$	$7.21 \cdot 10^{-9}$	$3.93 \cdot 10^{-9}$	**W14**	$2.58 \cdot 10^{-5}$	$2.07 \cdot 10^{-5}$	$1.23 \cdot 10^{-8}$	$9.00 \cdot 10^{-9}$
W5	$6.36 \cdot 10^{-6}$	$1.35 \cdot 10^{-6}$	$1.10 \cdot 10^{-8}$	$7.80 \cdot 10^{-9}$	**IND**	$9.80 \cdot 10^{-6}$	$4.78 \cdot 10^{-6}$	$1.32 \cdot 10^{-8}$	$9.87 \cdot 10^{-9}$
W6	$6.36 \cdot 10^{-6}$	$1.35 \cdot 10^{-6}$	$1.00 \cdot 10^{-8}$	$6.93 \cdot 10^{-9}$	**CHI**	$9.80 \cdot 10^{-6}$	$4.78 \cdot 10^{-6}$	$1.34 \cdot 10^{-8}$	$1.00 \cdot 10^{-8}$
W7	$1.28 \cdot 10^{-5}$	$7.77 \cdot 10^{-6}$	$9.18 \cdot 10^{-9}$	$5.93 \cdot 10^{-9}$	**JAP**	$9.82 \cdot 10^{-6}$	$4.78 \cdot 10^{-6}$	$1.35 \cdot 10^{-8}$	$1.02 \cdot 10^{-8}$
W8	$1.31 \cdot 10^{-5}$	$7.77 \cdot 10^{-6}$	$6.64 \cdot 10^{-9}$	$3.53 \cdot 10^{-9}$					

3.2. Effect factor of nano-CuO

It is preferred to use toxicity data (EC50s in three trophic levels) reflecting the entire ecosystem in *EF* calculation.[12] Nevertheless, the toxicity values of nano-CuO to the organisms living in sediment were rarely reported. In this study, one toxicity value (EC50 = 868 $\mu g \cdot g^{-1}$) of nano-CuO on an estuarine amphipod, *Leptocheirus plumulosus*,[7] was found and used in the calculation of sediment *EF* of nano-CuO. The sediment *EF* value of 7.025 $PAF \cdot m^3 \cdot kg^{-1}$ for nano-CuO was calculated and applied to calculate the freshwater sediment characterization factors of nano-CuO.

3.3. Characterization factors of nano-CuO

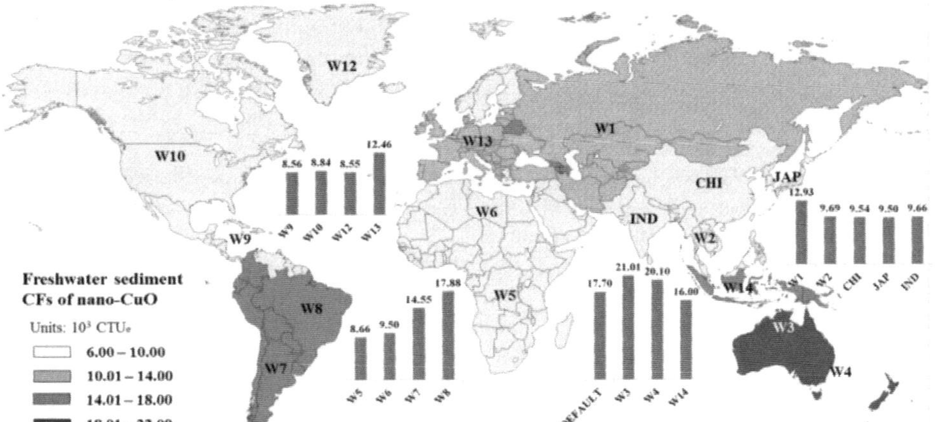

Figure 3. World map of freshwater sediment characterization factors (CFs) of nano-CuO, with histograms indicating the variation between the numerical values (units: 10^3 CTU_e) of 17 worldwide regions. The source of the underlying map was from Mapchart website.[13]

Based on the calculated *EF* and FFs of nano-CuO, the recommended world-wide nano-CuO sediment CFs were obtained and are depicted in figure 3. The deeper color on the map represents the larger *CF* of nano-CuO for the related region. The values reflect the potential influences of nano-CuO on organisms. The larger *CF* value of nano-CuO, the more sensitive is the ecosystem of the related region. These CFs can be used in the future LCA, when assessing products containing nano-CuO. This map can show a general view of nano-CuO sediment CFs in the world despite of the imprecision.

The sediment most likely to be affected by nano-CuO is in northern Australia (*CF* of $21.01 \cdot 10^3$ CTU_e, comparative toxic units) and the least likely is in northern Europe and northern Canada (*CF* of $8.55 \cdot 10^3$ CTU_e). A *CF* of $17.70 \cdot 10^3$ CTU_e is proposed for the DEFAULT region, which is a large value among all the studied regions. It means the impacts of nano-CuO to sediment would probably be overestimated for an unknown region. In addition, due to the relatively high uncertainty in the calculation of fate, exposure and effects of nano-CuO, the CFs proposed in this study should be classified as “indicative”.[9]

4. Perspectives

This work proposed the freshwater sediment characterization factors of nano-CuO in 17 subcontinental regions. Many limitations may affect the absolute values of the output: CFs of nano-CuO. The immature fate model of ENPs used in this study may bring uncertainties to FFs of nano-CuO. Moreover, *XF*, as a significant factor in *CF* calculation, may have big influences on the *CF* values. Thus, further investigation of *XF* is necessary. Furthermore, since the toxicity values of nano-CuO to the organisms living in sediment were rarely reported, just one toxicity value from literature was used in this study. The sediement *EF* of nano-CuO can be updated once more toxicity values are available. However, despite having defects, to the best of our knowledge, it is the first reports focusing on the sediment *CF* of ENPs. The method proposed in this study for calculating the sediment CFs of nano-CuO is promising. The recommended CFs for nano-CuO could also be used in the future when evaluating the ecosystem impacts of products containing nano-CuO.

Acknowledgement
Yubing Pu kindly thanks the Chinese Scholarship Council (CSC) for funding his PhD scholarship in France.

References

[1] Keller A A, McFerran S, Lazareva A and Suh S 2013 *J. Nanopart. Res.* **15**, 1-17.
[2] Sun T Y, Bornhoft N A, Hungerbuhler K and Nowack B 2016 *Environ. Sci. Technol.* **50**, 4701-11.
[3] Su Y, Yan X, Pu Y, Xiao F, Wang D and Yang M 2013 *Environ. Sci. Technol.* **47**, 4704-10.
[4] Wang Z, Zhang L, Zhao J and Xing B 2016 *Environ. Sci.: Nano* **3**, 240-55.
[5] Gottschalk F and Nowack B 2011 *J. Environ. Monit.* **13**, 1145-55.
[6] Pu Y, Tang F, Adam P M, Laratte B and Ionescu R E 2016 *Environ. Sci. Technol.* **50**, 9370-79.
[7] Hanna S K, Miller R J, Zhou D, Keller A A and Lenihan H S 2013 *Aquat. Toxicol.* **142-143**, 441-6.
[8] *ISO 14040: Environmental Management, Life Cycle Assessment, Principles and Framework*; International Standard Organization: Geneva, 2006.
[9] Fantke P, Bengoa X, Chappert B, Dong Y, Guignard C, Hauschild M Z, Huijbregts M, Jolliet O, Kounina A and Magaud V *USEtox® 2.0 Documentation (Version 1.00)*; USEtox® Team: 2015.
[10] Kasemets K, Ivask A, Dubourguier H C and Kahru A 2009 *Toxicol. In Vitro* **23**, 1116-22.
[11] Shaked S. Multi-continental multimedia model of pollutant intake and application to impacts of global emissions and globally traded goods. Ph.D. Dissertation, University of Michigan, Ann Arbor, MI, 2011.
[12] Hauschild M 2007 *Int. J. Life Cycle Assess.* **12**, 79-91.
[13] *Create Custom Map - mapchart* Website; http://mapchart.net/.

Phosphorus Sorption Capacities of Steel Slag in Pilot-Scale Constructed Wetlands for Treating Urban Runoff: Saturation Potential and Longevity

W J Guo[1, 2*], L Y Zhao [1], W H Zhao[1], Q Y Li[1] and Z B Wu [2]

[1]Basin Water Environment Research Department, Yangtze River Scientific Research Institute, Wuhan 430010, China
[2]Institute of Hydrobiology, Chinese Academy of Sciences, Wuhan 430072, China

*E-mail: guoweijie1986@163.com

Abstract. Two parallel pilot-scale integrated constructed wetland (ICW) systems were constructed on the bank of Nanfeihe River. The phosphate (PO_4^{3-}) isothermal adsorption properties of the upper substrate steel furnace slag (SFS) in up-flow chamber was investigated during one-year operation period. The maximum phosphorus (P) adsorption capacity of SFS 9, 11, 13, 15, 17, 19 months service time were 848.9 mg/kg, 968.1 mg/kg, 824.5 mg/kg, 788.7 mg/kg, 864.7 mg/kg and 960.3 mg/kg, respectively. The saturated adsorption amount of SFS had not decreased with the service time prolonging in ICW. The longevity of a full-scale system could not be reliably estimated only based on the theoretical saturated adsorption capacity from laboratory experiments.

1. Introduction

Constructed wetlands (CWs) have been accepted as a new ecological engineering technology all over the world and were more widely used in developing countries where sewage collection and treatment facilities are poorer[1].CWs have been proved to be an effective way to treat domestic sewage, industrial wastewater and livestock wastewater [2]. Furthermore, more studies and engineering practices have strongly approved that CWs could be used for non-point source pollution control [3,4,5].

It is generally believed that phosphorus (P) is one of the main control factors of water eutrophication that would be seriously exacerbated when the total P concentration in the water is greater than 0.02 mg/L [6]. It is of significantly important toalleviate water pollution and eutrophication by removing P pollutants effectively from the effluent. CWs have possessed many advantages of a low investment; convenient maintenance and outstanding ecological service function on removal and gradually become a new technology of wastewater treatment. In CWs, P in sewage was mainly removed by the adsorption and/or precipitation of substrates, bacterial activity, absorption of plant and algae and organic matter complexation [7]. The importance of substrates adsorption for Premoval by CWs has been recognized for nearly 3 decade.

However, it is worth noting that adsorption and/or precipitation of P by substrates in CWs is a finite process. Once the material becomes saturated, it will have to be either washed or replaced, or else P pollutants will be released and leads to water pollution [8].Therefore, the P adsorption capacity of substrates represents a central parameter for comparing and selecting candidate materialsto be used as key removal media in CWs [8,9]. Nevertheless, while acquiring the information of adsorption capacity of some candidate substrates can be used to estimate the maximum amount of P from the influent, it is difficult toprovide an accurate estimation of the longevity of CWs. The longevity of CWs is restricted by the P saturation potential of the substrates as well asthe feeding regime (e.g. hydraulic retention time, HRT), plant growth, microorganism activity and P concentrations of influent. Thus it is obvious that using the Langmuir equation with experimental data of laboratory test cannot obtain the realistic adsorption capacity of substrates in pilot-scale CWs and thelongevity. It was recommended that P batch experiments should be coupled with a longer-term investigation of substrates performance through columns or even pilot-scale study for the better estimation of P removal efficiencies and retention capacities of different substrates. Therefore, it is necessary to carried out the study of the saturated adsorption capacity and variation tendency of substrates with the service time prolonging during based on pilot-scale CWs, which is of great benefit to predict the long-term performance and improve the purification capability of CWs.

In this study, two parallel pilot-scale integrated constructed wetland (ICW) systems were constructedon the bank of Nanfeihe River, a seriously polluted urban river, for purifying urban runoff. The detail operation and pollution removal efficiencies were performed in another paper[10]; here, we focused on the phosphate (PO_4^{3-}) isothermal adsorption properties (linear Langmuir equation) of the upper substrate Steel furnace slag (SFS) in up-flow chamber based the ICW in an operation period of one year. The maximum P adsorption capacity of SFS was also investigated with its service time prolonging in ICW. This long-term pilot-scale study was conducted to provide a theoretical basis for optimizing the purification performance and realizing the long-term operation of CWs.

2. Materials and methods

2.1. Site description and design of the ICW

Two parallel pilot-scale ICW systems (each was composed of a IVCW and HSSF-CW in series) were constructed on the bank of Nanfeihe River, a tributary and the biggest pollution source of Chaohu Lake, in Heifei, China. The ICW systems were constructed in early May 2011. Each was designed as a combination system of a down-flow chamber (DFC, 50 m^2 and planted with *Canna indica*), an up-flow chamber, and a horizontal subsurface-flow chamber in series (Fig.1) (For more information, refer to Guo *et al.* [10]. This study mainly focused on the IVCW, especially the DFC, to explorethe saturated adsorption capacity and variation tendency of SFS used in the chamber.

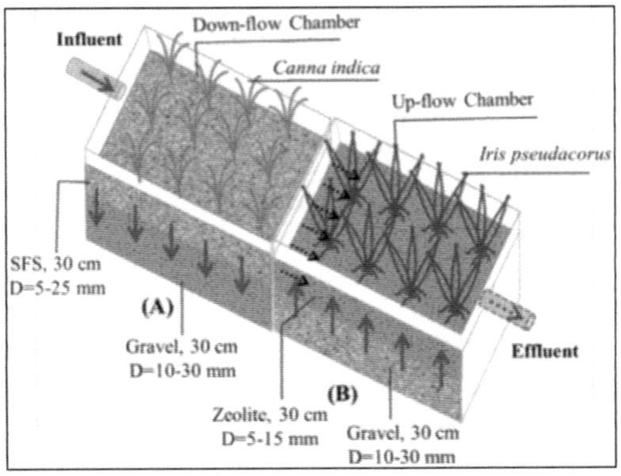

Figure1. The down-flow chamber(A) and up-flow chamber(B) of the IVCW.

2.2. Operation conditions

The wastewater was pumped from the sedimentation tank in a pump station and induced in each ICW unit per day to yield a hydraulic loading rate of 100 mm/d, and the theoretical HRT was about 2.2 d [10].

2.3. Sampling and analysis

Five representative rhizosphere samples were collectedfrom five target plants in DFC and homogenized well to makea composite sample once in every two months frommid-March 2012 to mid-January 2013. After all litter was removed, the field-moist samples were sieved with 10 mm separating sieve and dried at room temperature, then sealed in a desiccators for determination.

The maximum P adsorption capacity was determined using a slightly modified batch equilibrium technique as described by Drizo *et al.* [8,9]. The maximum P adsorption capacity was calculated from the difference between the initial P concentration added to the solution and the P concentration in the supernatant, using the nonlinear form of the Langmuir equation:

$$C_e/q= C_e /b+1/ab \quad (1)$$

where C_e is the concentration of P in the solution at equilibrium (mg/L), q is the amount of P adsorbed per unit mass of material (mg/g), b is the maximum P adsorption capacity (mg/g), and a is the constant related to the binding strength of P onto the material.

2.4. Statistical analysis

Statistical analyses were performed using SPSS13.0 for Windows (SPSS Inc., Chicago, IL, USA;Version 13.0). The experiment data were presented asmean ± standard deviation. Comparison of the averages was analyzed using one-way ANOVA followed by LSD tests.

3. Results and Discussion

In the research of adsorption kinetics, Langmuir formula and Freundlich formula, as two common isotherm equations, are often used to represent the adsorption process of different materials on pollutants. Wherein, Langmuir equation is constructed by Langmuir in 1961, and the original form is $q=abC_e/(1+aC_e)$. Several important parameters about adsorption characteristics, including a, q and b (see above), can be obtained according to this equation consequently that has been widely used in characterizing the adsorption processes of different medium on pollutants. When $a>0$, the reaction between materials and pollutants at the same temperature conditions can occur spontaneously, and higher values indicate a stronger spontaneity degree and adsorption capacity on pollutants and more stable products.

If pollutant concentrations (i.e. initial P concentration in experiment) are relatively higher, the theoretical value calculated from fitting Langmuir equation could differ from the true value to a certain extent. Whereas, when the pollutant concentration is relatively lower, the equation could be capable of fitting the actual adsorption process [11].Therefore, the maximum PO_4^{3-}concentration (concentration gradient of KH_2PO_4 solution) in this study was set to 80 mg/L, simulating the actual adsorption process of the SFS as accurately as possible.

In the study, the Langmuir isotherm plot of P adsorption data for SFS collected from different dates (i.e. different service time) were shown in Figure 2. All the values of the correlation coefficients (r) estimated by Langmuir isotherm plot of SFS from different dates were greater than 0.9 and present the better fitting goodness. Therefore, the Langmuir model is suitable for simulating the isotherm plot of P adsorption for SFS of different service time.

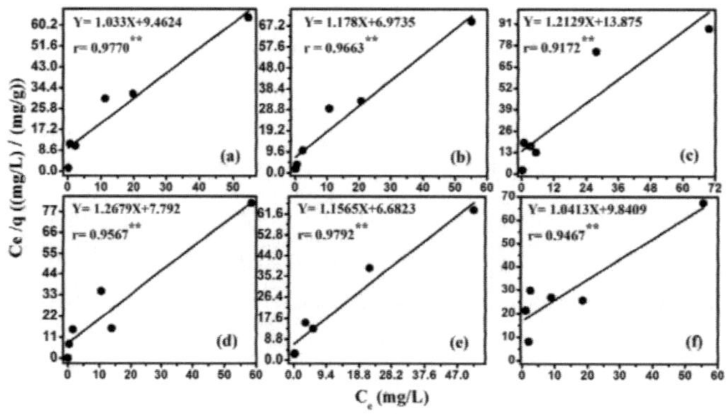

Figure 2. Langmuir isotherm plot of P adsorption data for SFS collected from different dates: (a) March, (b) May, (c) July, (d) September, (e) November 2012 and (f) January 2013. The derivative of the slope of the regression line was the maximum apparent P adsorption capacity. (*and** denote significance at 0.05 and 0.01 levels, respectively).

The maximum P adsorption capacity (b) and binding energy constants (a) as estimated by Langmuir isotherm plot of SFS from different dates were illustrated in Table 1. The b value of SFS from March 2012 (9 months of service time), May 2012 (11 months of service time), July 2012 (13 months of service time), September 2012 (15 months of service time), November2012 (17 months of service time), January 2013 (19 months of service time) were 848.9 mg/kg, 968.1 mg/kg, 824.5 mg/kg,

788.7 mg/kg, 864.7 mg/kg and 960.3 mg/kg, respectively. Based on these results, it is concluded that the saturated adsorption amount of SFS have not decreased with the prolonging of service time in the pilot-scale ICW.

Table1. Maximum P adsorption capacity (b), binding energy constants (a) and correlation coefficients (r) as estimated by Langmuir isotherm plot of SFS from different dates.

SFS of different service time	Maximum P adsorption capacity,b (mg/kg)	binding energy constants, a	correlation coefficients, r
Mar. 2012	848.90	0.17	0.9770**
May. 2012	968.05	0.11	0.9663**
Jul. 2012	824.47	0.09	0.9172**
Sep. 2012	788.71	0.16	0.9567**
Nov. 2012	864.68	0.17	0.9792**
Jan. 2013	960.34	0.11	0.9467**

According to maximum P adsorption capacity obtained in the experiment, it could be expected to make a preliminary estimate forthe life expectancy of SFSin ICW. The working life and durability of CWs is closely related to the concentration and type of pollutants in the influent.

Assuming the CWs possess a same removal rate whether under the high hydraulic loading rate (HLR) or low HLR, the theory service life of CWs could be calculated as follows. It is assumed that 2.3 g P per person per day were discharged in municipal wastewater. According to the wetland design guidelines (EC/EWPCA) described by Drizo *et al.*[9], a 3 m^3 wetland (wetland area is 5 m^2 and depth is 0.6 m) was required to constructed to purify the discharged sewage andabout 4.5 tons SFS needed to be filled in the CWs. Applying the saturated adsorption capacity (the average value was 0.88 kg P, see Table 1) to the 3 m^3 wetland mentioned above, 4.5 ton SFS could remove 4.0 kg P and the life expectancy was about 9 years.

In CWs, P in the influent can make complexation with Fe, Al and Ca ion and be removed via adsorption and/or precipitation by substrates, which have been considered as the main way and mechanism for P removal [12]. The P maximum adsorption capacity and the longevity of substrates are two important parameters in the practical application of CWs. It was shown that the saturated adsorption capacity of SFS had not continued to decrease with the increase of its service time in wetland system after nearly a year research. The P adsorbed on the SFS surface can be removed to some extent by the absorption of plants, which may be one of the reasons for the substrates maintaining the adsorption longevity in CWs [13]. Furthermore, the ICW system was operated in an intermittent way and batch injection of the influent can maintain the environment in which the anaerobic and oxygen enrichment condition occurred alternately [14]. In this study, the P adsorption ability of SFS had been recovered to a certain degree at least in the short term.

4. Conclusions

In conclusion, the Langmuir model could be suitable for simulating the phosphorus adsorption isotherm of SFS, the maximum adsorption capacity of whichhad not continued to decrease with the service time increase in ICW during operating period. Consequently, the longevity of a full-scale system could not be reliably estimated only based on the theoretical saturated adsorption capacity obtained by laboratory experiments. Besides, it was unreliable to take itas a central parameter for comparing and selecting candidate materials to be used as P removal media in CWs. The longevity of full-scale CWs should more reliably be evaluated from a P saturation value derived from a long-term column or pilot-scale study and that could provide useful data about the suitability of substrates in CWS for long-term P removal and the prediction of the longevity of the pilot-scale design.

Acknowledgement
The research was financially supported by National Natural Science Foundation of China (Grant No. 51509011) and Special fund for basic scientific research business of central public research institutes (Grant No. CKSF2016002/SH).

References

[1] Hadad H R, Maine M A and Bonetto C A 2006 Macrophyte growth in a pilot-scale constructed wetland for industrial wastewater treatment. *Chemosphere* 63(10): 1744-53.

[2] Cheng S, Grosse W, Karrenbrock F and Thoennessen M 2002 Efficiency of constructed wetlands in decontamination of water polluted by heavy metals. *Ecol. Eng.* 18(3): 317-325.

[3] Lai D Y F and Lam K C 2009 Phosphorus sorption by sediments in a subtropical constructed wetland receiving stormwater runoff. *Ecol. Eng.*35(5): 735-743

[4] Wu C Y, Kao C M, Lin C E, Chen C W and Lai Y C 2010 Using a constructed wetland for non-point source pollution control and river water quality purification: a case study in Taiwan. *Water Sci. Technol.*, 61(10): 2549-55.

[5] Maniquiz M C, Lee S Y, Choi J Y, Jeong S M and Kim L H 2012 Treatment performance of a constructed wetland during storm and non-storm events in Korea. *Water Sci. Technol.*,65(1): 119-126.

[6] Sallade Y E and Sims J T 1997 Phosphorus transformations in the sediments of Delaware's agricultural drainage ways: II. Effect of reducing conditions on phosphorus release. *Environ. Qual.* 26(6): 1579-88.

[7] Lantzke I R, Mitchell D S, Heritage A D and Sharma K P 1999 A model of factors controlling orthophosphate removal in planted vertical flow wetlands. *Ecol. Eng.*12(1): 93-105.

[8] Drizo A, Comeau Y, Forget C and Chapuis, R P 2002 Phosphorus saturation potential: a parameter for estimating the longevity of constructed wetland systems. *Environ. Sci. Technol.* 36(21): 4642-48.

[9] Drizo A, Frost C A, Grace J and Smith K A 1999 Physico-chemical screening of phosphate-removing substrates for use in constructed wetland systems. *Water Res.* 33 (17), 3595-3602.

[10] Guo W J, LiZ, ChengSP, LiangW and WuZB 2014 Performanceof a pilot-scale constructed wetland for stormwater runoff and domestic sewage treatment onthe bank of a polluted urban river, *Water Sci. Technol.*69 (7): 1410-18.

[11] Li Q N, Zhang J Q, Xie J and Xu W L 2011 Performances and dynamics of substrate in constructed wetland system adsorbing phosphorus. *Tech. Water Treat.* 37(9): 64-67.

[12] Johansson L 1999 Blast furnace slag as phosphorus sorbents-column studies. *Sci.Total Environ.* 229(1): 89-97.

[13] Greenway M and Woolley A 1999 Constructed wetlands in Queensland: performance efficiency and nutrient bioaccumulation. *Ecol. Eng.* 12(1), 39-55.

[14] Shilton A, Pratt S, Drizo A, Mahmood B, Banker S, Billings L, Glenny S and Luo D 2005 Active filters for upgrading phosphorus removal from pond systems. *Water Sci. Technol.*51(12): 111-116.

Study on Fluoride Removal from Aqueous Solutions by a Novel Defluoridation Adsorbent NEWDEF®

Chi Zhu[1, 2], Shui Wang[1], An-juanCai[1], Dong Wang[1], Wei-Xia Wang[1], Jia-ming Zhu[3] and Kai Nong[3]

[1]Jiangsu Provincial Academy of Environmental Science, Jiangsu provincial key laboratory of environmental engineering, Nanjing 210036, China.
[2]State Key Laboratory of Lake and Environmental Sciences, Nanjing Institute of Geography and Limnology, Chinese Academy of Sciences, Nanjing 210008, China.[3]ChinaSchool of Chemistry Life Science,Nanjing University Jinling College,Na njing 210089,China.

zhuchi126@126.com

Abstract. Batch experiment was carried out to determine the effectiveness of the novel composite defluorinating agent NEWDEF® for fluoride removal from aqueous solution. The results showed that the maximum adsorption capacities for F^- were 4.09mg·g^{-1}. Kinetics data obtained for the adsorption process fitted the Lagergren First-order equation. The adsorption kinetics and isotherm were found well fitted by Langmuir isotherm models and a pseudo-second-order kinetic, showing F^- removal belonged to monolayer adsorption process. The reaction of F^- adsorption from aqueous solutions by NEWDEF® was an endothermic process. Micro-columnand home-used equipment defluoridation test indicated that NEWDEF® can be perfectly suitable defluorinating agent in F^- removal for drinking water in small towns.

1. Introduction

Because of a shortage of freshwater, it's the major, important and preferable source of drinking water in developing countries especially, while ground water only makes up 0.6% of the total water resources on the earth [1]. Fluoride removal from drinking water has become an urgent task all around the world, particularly in China, owing to the wide distribution and high concentration in groundwater. There are more than 20 nations which are suffering from endemic fluorosis, for example, China, Africa, India, Spanish and the Middle East in particular [2]. In China, which has a large area with fluoride, fluorosis were found prevalently in 29 provinces, municipalities and autonomous regions. According to the investigation results by the Ministry of Health and the Ministry of Water resources in China, there were 81.41million people suffered from fluorosis. There were about 100million people drinking water with fluorine content had over 1.0mg/L and approximately5million people whose potable water with fluorine exceed 5.0mg/L [3].Fluoride is one of the necessary microelement for human body. However, its excessive intake could be harmful to human health, which may result in detrimental physical effects include dental fluorosis, skeletal fluorosis etc, such as hypersensitivity reactions and kidney disease.

The mainly methods to remove excess fluoride from water including ion-exchange, adsorption, chemical precipitation, electro dialysis, ultra filtration, reverse osmosis, membrane separation treatment and so on. Adsorption was one of the most extensive application and the commercially viable technology. The commonly used defluoride adsorbents include alumina, bone char, zeolite, gypsum material, kaolinite, cement clay, montmorillonite and rare earth metal [4-8]. However, the adsorbents have been restricted due to their unfeasibility, inoperability, expensive price, unsafety or not widely adaptability. Therefore, a novel adsorbent with high efficiency, practically, economical and safety was developed for defluoridation in our study.

2. Materials and methods

2.1. Materials

NEWDEF® is a new high efficiency drinking-water defluoridation media, which was natural mineral and insoluble in water. The particle size of the media was 0.5mm-1.5mm and bulk density was 1.33g/ml.

2.2. Methods

2.2.1. Batch adsorption experiments. NEWDEF® was sieved firstly, and then agitated with 1.0 g in 100 ml beaker of fluoride aqueous solution of desired concentration in triplicate by stirring with speed of 150 rpm. At the end of predetermined time intervals, NEWDEF® defluorinating agent was removed from the solutions by centrifugation at 10000 rpm for 5 min, and finally determining the residual concentration of F^-. The adsorption capacity (q) and the percentage removal (%)by NEWDEF® were calculated by using the following equation:

$$q = \frac{(Ci - Ce) \times V}{1.0} \quad (1)$$

$$\text{The percentage removal (\%)} = \frac{(Ci - Ce)}{Ci} \times 100\% \quad (2)$$

Where q (in mg/g) was capacity, the Ci and Ce were the initial and equilibrium concentrations of F^- in aqueous solution (mg/L), V was the solution volume(L).

2.2.2. Micro- column experiments. 30 g NEWDEF®, 30 g activated alumina, 30 g activated zeolite was added to three 50ml acid burettes respectively, with 2ml/min-2.5ml/min flow of raw water. Then the fluoride content was determined until more than 1.0mg/L. The fluoride content of raw water was 2.4mg/L, which was from a water plant in Puyang City in China.

2.2.3. Home-used defluoridatione quipments. 30 kg filter were filled in the household defluoridation equipment with 2.0 L/min flow of fluoride water, and the water fluoride content was determined until the fluoride content over 1.0 mg/L. Then the flow rate mode was changed to 1 hour per day intermitted flow, the effluent fluoride concentration was measured and the flow volume was recorded simultaneously.

2.2.4. Statistical analysis. Descriptive statistics, analysis of variance and treatment group comparisons were calculated using the computer program, GraphpadPrizm (Version4) at 0.05 and 0.01 significance level.

3. Results

3.1. Effect of particle size and contact time on fluoride removal

The effect of particle size and contact time was conducted by the experiments on the fluoride removal. The NEWDEF® adsorbent media size was range from at 0.2-0.5, 0.5-1.0 and 1.0-1.5mm diameter, and the result was displayed in Fig.1. It is evident that there was a sharply rising in fluoride removal in the initial stages. The effect of particle size on the defluorinating efficiencies was limited, of which maybe the smaller particle sizes has the relatively higher external surface area. But the efficiency of F^- removal was no significantly different ($P>0.05$) as particle size increase, that showed particle size has no obvious effect on F^- removal.

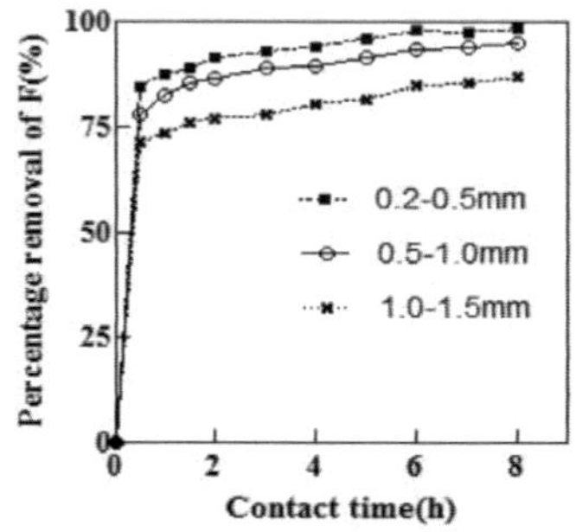

Fig.1. Effect of particle size and contact time on fluoride removal

3.2. Determination of adsorption kinetic constants of NEWDEF® in defluorinating process

The adsorption kinetic was determined by using Lagergren first-order rate at room temperature 25℃ (±1℃), and the rate constants of adsorption (K_{ad}) by NEWDEF® were obtain by the following equation expression [9]:

$$\log(qe - qt) = \log qe - (\frac{K_{ad}}{2.303})t \qquad (3)$$

Where qe and qt (both in mg/g) were the amount of F^- adsorbed at equilibrium and any time't', respectively. K_{ad} (h^{-1}) was the kinetic rate constants of adsorption. The value of k_{ad} was determined by 'log (*qe-qt*)'versus time as shown in the graph. The correlation coefficient of Lagergren's plot for kinetic modeling was 0.9902 (Fig.2). The results reveal that the adsorption process fitted the Lagergren kinetic modeling better and the rate constant of adsorption was 14.28 h^{-1} for F^-.

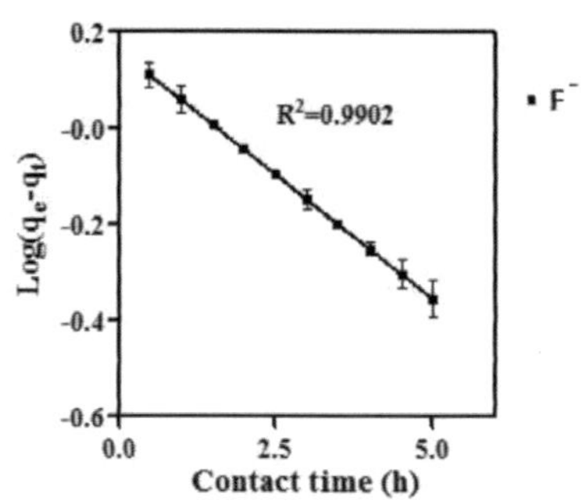

Fig.2. Lagergren's plot for kinetic modeling for fluoride removal

3.3. Adsorption isotherm

The distribution of F^- in aqueous solution by NEWDEF® were determined by Freundlich or Langmuir, which indicate the distribution equilibrium of F^- between the solid phase or the liquid phase in the adsorption process.

The Langmuir adsorption isotherm is expressed in a linear form, which demonstrated a monolayer sorption. The Langmuir adsorption assumes that maximum adsorption occurs when the surface is covered by defluorinating agents, and the adsorption onto a surface with identical sites of the media.

The equations are given as following:

$$qe = \frac{abCe}{1 + bCe} \tag{4}$$

$$\frac{Ce}{qe} = \frac{1}{a}Ce + \frac{1}{ab} \tag{5}$$

The Langmuir and Freundlich adsorption isotherms of F^-adsorption were displayed in Fig.3. From the graph (*Ce*/*qe*) vs. *Ce* showed a straight line and the empirical constant could be obtained from the kinetic data. Which '*Ce*' (mg/L) represents the equilibrium concentration, '*qe*' (mg/g) represents the amount adsorbent at equilibrium, 'a' represents the constant related to adsorption capacity, and 'b' is the constant about the energy of adsorption.

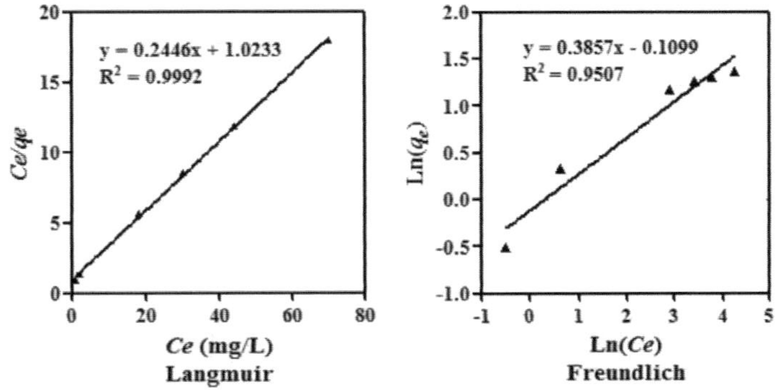

Fig.3. The Langmuir and Freundlich adsorption isotherms of F^-adsorption

From the correlation coefficient of both equations with the experimental data (as shown in Fig.3), it was found that Langmuir adsorption isotherm ($R^2_{F^-}$=0.9992) fitted the experimental data better than that of the Freundlich adsorption isotherm ($R^2_{F^-}$=0.9507). The adsorption capacity of NEWDEF® was 4.09mg/g.

3.4. Micro- column fluoride removal experiment

The results in Table 1 and Fig. 4 illustrate the treated water was only 0.97L and 2.23L by activated zeolite and the activated alumina respectively, while treatment by NEWDEF® was 11.35 L of raw water. The amount of water treated by NEWDEF® is 11.7 times that of activated zeolite and 5.09 times that of activated alumina in the same condition.

Table 1 Flow testing results by micro-column defluoridation

Material / Flow rate	NEWDEF® (30g)	Activated alumina (30g)	Activated zeolite (30g)
43ml/min (Equivalent to a ton of material 86m³/h effluent flow)	Fluorine content in effluent	Fluorine content in effluent	Fluorine content in effluent
	0.20mg/L	0.90mg/L	1.0mg/L

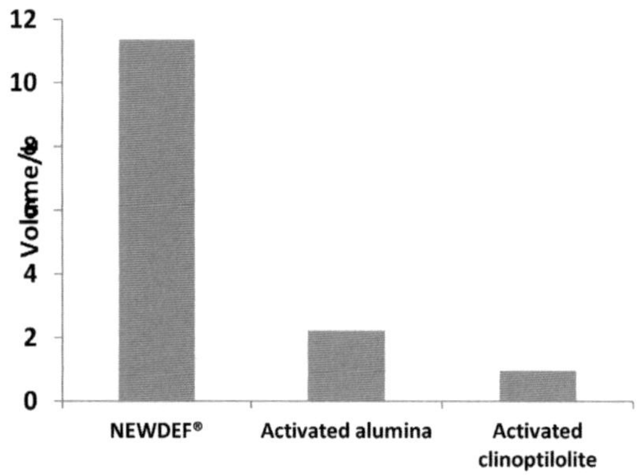

Fig.4. Thevolume of water treated by different media under the same condition

3.5. Removal of F⁻ by home-used defluoridation equipment

The results from Fig. 5 showed that NEWDEF® filter in the home-used equipment can still maintain the same fluoride removal capacity as the micro-column test. It can still restore part of the fluoride effect by replace continuously flow with intermitted flow as the material saturated with the raw water, and NEWDEF® can maintain a more stable effect.

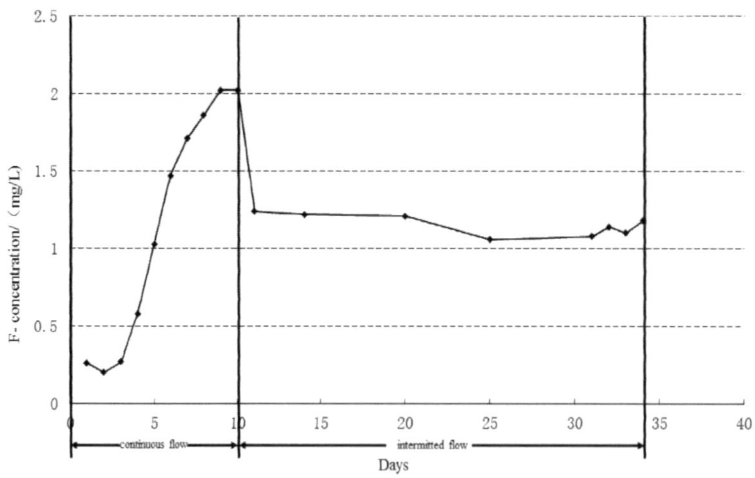

Fig.5. Thefluoride removal by home-used equipment

4. Conclusions

The effectiveness and capacity of defluorinating agent NEWDEF® was investigated, and the conclusions are made as follows:

(1) The adsorption process by NEWDEF® attain equilibrium in 6 h, and F^- removal was no significantly different ($P>0.05$) as particle size increase.

(2) The removal of F^- by NEWDEF® was found to fit Langmuir adsorption isotherm models ($R^2=0.9992$).

(3) The adsorption kinetics data indicated a pseudo-second-order kinetic, showing F^- removal belonged to monolayer adsorption process and the reaction of F^- adsorption by NEWDEF® was an endothermic process.

(4) Micro-column and home-used defluoridation tests indicated NEWDEF® is 11.7 times that of activated zeolite and 5.09 times that of activated alumina in removal F- and can maintain a more stable effect.

Acknowledgements

This work was supported by Water Pollution Control and Treatment of National Science and Technology Major Project (Grant No. 2012ZX07506-001) and the National Natural Science Foundation of China (Grant No. 31300462) and Jiangsu Planned Projects for Postdoctoral Research Funds (Grant No. 1302003C).

References

[1] Sharma B S, Agrawal J and Gupta A K 2011 Asian J. Environ. Biol. Sci. 2 p 131-134
[2] Meenakshi and Maheshwari R C 2006 *J. Hazard. Mater.* 137(1) p 456-463
[3] Zhang Y, Xie H, Jia Wand Wu Z 2005 *J. Wuhan Univ. (Nat. Sci. Ed.)* 51(4) p 457-460 in Chinese
[4] Kang W H, Kim E I and Park J Y 2007 *Desalination* 202 p 38-44
[5] Ma W, Ya F Q, Han M and Wang R 2007 *J Hazard Mater* 143(1-2) p 296-302
[6] Natrayasamy V and Meenakshi S, 2008 *J Fluorine Chem* 129 p 503-509
[7] Cheng X, Zhong Y, Wang J, Guo J, Huang Q and Long Y 2005 *Microporous Mesoporous Mater* 83 p 233-243
[8] Islam M and Patel R K 2007 *J Hazard Mater* 143 p 303-310
[9] Naseem Rauf and Tahir S S 2000 *J Chem Therm* 32 p 651-658

Dormant Progagule Banks of Cladoceran Communites and Hatching Success from the Sedimentin China

Chi Zhu[1,2], Shui Wang[1] and Fan-xiang Kong[2]

[1]Jiangsu Provincial Academy of Environmental Science, Jiangsu provincial key laboratory of environmental engineering, Nanjing 210036, China.
[2]State Key Laboratory of Lake and Environmental Sciences, Nanjing Institute of Geography and Limnology, Chinese Academy of Sciences, Nanjing 210008, China

zhuchi126@126.com

Abstract: The resources of cladoceran communitesin lakes in typical eutrophic shallow lakes in Chinawere investigated. Diapausing eggs of cladocerans were collected from the surface sediment of typical eutrophic shallow lakes and then hatched in the laboratory. The most common and dominant genera were *Bosmina* sp., *Ceriodaphnia* sp.,*Alona* sp.and*Moina* sp.The density of large cladocerans in the hatchlings was lower than the small cladocerans (Bosmina, Ceriodaphnia,Alonaand Moina). The distribution and abundance and species of Daphnia community were significantly influenced by TN and fish community.

1. Introduction

A portion of thesediapausing eggs will experiencea period of dormancy and then hatch as long as the climate gets favorable, while the otherdiapausing eggs remained will constitute the growing diapausing eggs library [1]. The diapausing eggs accumulated in this library was produced in different periods by various zooplankton animals that lived in different habitats and occupied different ecological niches, some of these eggs still have incubation capacity after decades or even hundreds of years of their production [2]. The research of diapausing eggs and diapausing egg library offers important applications in several fields,such as natural protection, diversity exploration, paleolimnology, taxonomy, evolutionary ecology, ecological biogeography and population and community ecology [3-4].

Previous studies have investigated the zooplankton communities in the thirty lakes in the middle and lower Yangtze River, finding that the density of large species of Daphnia is generally small, while small species of *Moina micrura* (*M. micrura*), *Diaphanosoma brachyurum* (*D. brachyurum*) and *B. coregoni*were the most common and dominant species [5]. In addition, Gao and Xu (2009), Wang

(2003) and Zheng (2009) [6-8] investigated Poyang Lake, Ou Jiang and Dianshan Lake respectively, showing that Daphnia is no longer the dominant species and that large zooplankton community has been declining obviously. Therefore, we studied the dormant crustaceancommunity from typical eutrophic shallow lakes in China. The aims of our work were to identify the eggs and investigate the hatching success, whichcould provide the basic informationfor large zooplankton community diversity in typical eutrophic shallow lakes in China.

2. Material and methods

2.1. Study site

Limnological characteristics of the studied lakes were shown as Table 1.

Table 1 Limnological characteristics of the studied lakes in China

Lake	Latitude	Longitude	Altitude(m)	Samples	Area(km²)	Depth(m)
Dianchi	24°45′~24°57′	102°38′~102°42′	1887	7	309	5.3
Xingyun	24°18′~24°21′	102°45′~102°47′	1722	5	38.6	5.6
Chao	31°28′~31	117°20′~11	23	10	753	3.06

2.2. Sampling and sample processing.

Qualitative zooplankton samples were collected using a conical tow-net (diameter, 64 μm mesh), combining vertical hauls with horizontal hauls. Samples were washed in the net and fixed in 100% ethanol.

Diapausing eggs of zooplankton were collected within the surface sediments (0-5 cm), which were sampled using a Petersen grab to ensure that only the top-most layer of flocculent mud was taken. These samples were stored without fixation. Subsamples of surface sediment were firstly washed through a 1.25mm mesh metal sieve and then through a 200 μm mesh metal sieve. Subsequently, the residue was packaged respectively in dark at 4℃ in a week for later use.After the cold storage, quantitatively take out the mud sample to a 250ml beaker, add 200ml filtrated lake water (using 0.45μm filter membrane), and make hatch in the light incubator. Incubation experiment went on until no new individual hatchlings came. Incubation temperature was 24 ± 2 ℃, 40μmol/m^2·s and darkness14L: 10D[9-10].In the laboratory, hatchlings in the zooplankton samples were identified using a compound microscope with reference to Jiang and Du[11] and Shen [12] respectively.

Total phosphorus (TP), total nitrogen (TN) and sediment total organic carbon (STOC) were measured. The way of determination of total phosphorus in the sediment was called SMT [13]. Extraction of total nitrogen used the method of the persulfate digestion [14], and determination of total nitrogen in digestion solution was spectrophotometry. The column of total organic matter was supposed to be determined by the potassium dichromate method [15].

2.3. Statistical analysis.

Multivariate ordination techniques (using Cladoceran diapausing eggs abundance) were carried out using the software CANOCO 4.5[16].Descriptive statistics and Pearson and Spearman correlations for

IOP Publishing
doi:10.1088/1755-1315/51/1/012023

environmental variables and biologic variations about hatchlings were calculated using SPSS (version 16.0) at 0.05 and 0.01 significance level.

3. Results

Diapausing eggs of cladocerans were collected from the surface sediment of the eutrophic shallow lakes and hatchling results were shown in Fig.1-Fig.2. Sediment incubation from Xingyun Lake were mainly dominated by copepods with the density of 39~422/m^2.There is only one site in Xingyun lake which hatchlings density of Ceriodaphnia quadrangula (Ceriodaphnia quadrangula) is up to 344 /m^2; the hatchlings density in other sampling sites were just 101 ± 15 /m^2. For the incubation of diapausing eggs in Dianchi, large cladocerans Daphnia hyalina and Daphnia pulex were detected in two sites (S-5# and S-7#).The most notable thing was the density of D. pulex from sediment incubation could reach up to 336/m^2 in the south of Dianchi lake near the Haikou site (S-7#).

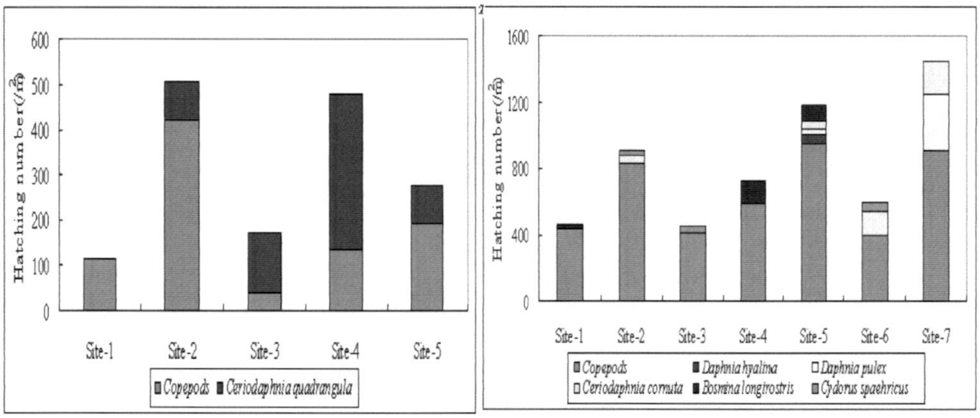

Fig. 1 Taxonomical composition of the resting egg collected in the sediment in XingYun Lake and Dianchi Lake (Left: Xingyun Lake; Right: Dianchi Lake)

The hatchlings of chaohu were shown in Fig2. Daphnia has been found in the sediment, but the dominant genera were small cladocerans. In Chaohu, there were five species of Daphnia in sediment hatchings, including *D.longispina*, *D. pulex*, *D. cucullata*, *D. hyalina* and *D.magna*. The small Cladocerans were dominant, in which *Bosmina* and *Daphniidae* were the most common species and accounted for about 20% ~ 30% of the hatchlings.

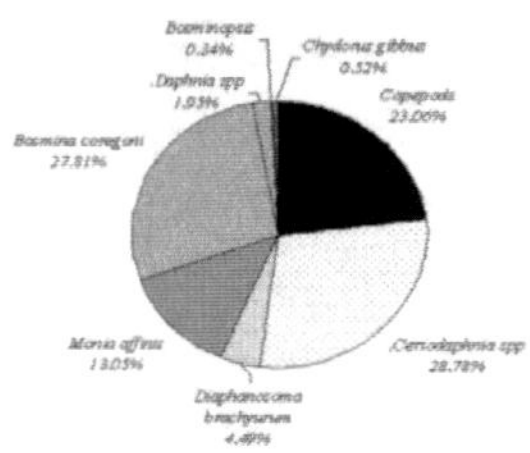

Fig. 2 Composition (%) of main species hatched from the sediment in Chaohu Lake

4. Dissusion

In this study, TN was significantly positively correlated with the distribution and species of the diapausing eggs by Pearson correlation coefficient sand probability levels between parameters of hatchlings and physico-chemical factors in lakes. The results indicated that TN may influence the abundance of the eggs in sediments.

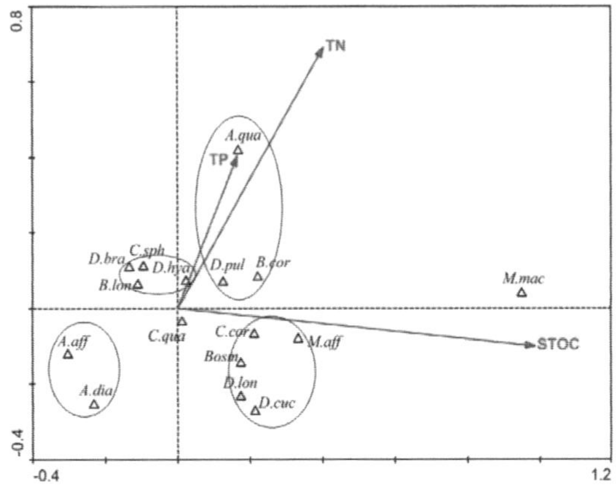

Fig. 3 CCA ordination diagram of 16 species of the Cladoceran

The CCA first ordination axis was significantly positively correlated with STOC, indicating that STOC was the mainly environmental factor in Cladoceran species distribution. According to the results of CCA, 16 Cladoceran species were divided into four groups (Fig. 3). *Alona affinis* and *Alona diaphana* were classified as group1, which was adaptive to low TN, TP and TOC environment. *Diaphanosoma brachyurum*, *Bosmina longirostris*, *Chydorus sphaericus*, *Daphnia hyaline* were classified as group 2, which lives well in the environment with high TN and TP but low STOC. *Alona quadrangularis*, *Daphnia pulex* and *Bosmina coregoni* as group 3 was adaptive to the environment with high TN, TP and TOC. The rest five species classified as group 4 was adaptive to the environment with high STOC but low TN and TP.

P. M. Mara-Barbosa *et al.* [10]found a large number of diapausing eggs produced by cladoceran and rotifer in the sediment of eutrophic lake, with the abundance ranging from $203.2\times10^3/m^2$ to $80\times10^3/m^2$.This may be caused by the fact that the distribution and incubation of diapausing eggs of cladoceran community in natural water body were easily affected by the biological disturbance and ecological restoration, for example, dredging, intercepting and renewing water.

In recent years, highly planktivorous fish stock along the middle and lower reaches of Yangtze River, the herbivorous zooplankton assemblage is characterized by small species and low density of large species. Fish communities from lakes in south western China were mainly composed of autochthonous fish species, accounting for 68%. The fish species composition and productions in southwest Lakes were different from those in the middle and lower reaches of Yangtze River, so they could affect the zooplankton in different ways, including size-structure of the community and the diapausing eggs distribution and abundance.

Acknowledgments
This work was supported by Science and technology support project of Jiangsu Province "Construction on comprehensive management platform of the typical regional ecoenvironmental in Jiangsu Province of

China" and the National Natural Science Foundation of China (Grant No. 31300462) and Jiangsu Planned Projects for Postdoctoral Research Funds (Grant No. 1302003C).

References

[1] Brendonck L, Meester LD 2003 Egg banks in freshwater zooplankton: evolutionary and ecological archives in the sediment*Hydrobiologia*491 p 65-84

[2] Charles E. King and Manuel Serra 1998 Seasonal variation as a determinant of population structure in rotifers reproducing by cyclical parthenogenesis *Hydrobiologia*387/388 p 361-372

[3] Brendonck L, Meester LD 2003 Egg banks in freshwater zooplankton: evolutionary and ecological archives in the sediment *Hydrobiologia* 491 p 65-84

[4] Vandekerkhove J, Brenda Niessen, Declerck S, Jeppesen E, José María Conde Porcuna, Brendonck L, Luc De Meester 2004 Hatching rate and hatching success with and without isolation of zooplankton resting stages *Hydrobiologia* 526 p 235-241

[5] Songbo Wang, Ping Xie, Shikai Wu, Xiaomin Liang 2007 Phytoplankton biomass in relation to nutrients and zooplankton in thirty subtropical lakes adjacent to the Yangtze River, China *Fundamental and Applied Limnology*169/1 p 49-55

[6] Gao Qian, Xu Zhao li 2009 Species composition and diversity of zooplankton in Oujiang River estuary in summer and autumn*Chinese Journal of Ecology* 28(10) p2048-2055

[7] Zheng Xiao yan, Wang Liqing, Gai Jianjun, Ji Gaohua, Fan Zhifeng, Zhang Ruilei 2009 Dynamic and Community Structure of Zooplankton in Dianshan Lake*Chinese Journal of Zoology* 44(5) p 78-85

[8] Wang Jinqiu, Wu Jianping, Yu Yanbin, Wang Tianyu, Cheng Gong 2003 The specific list, Quantitative Distribution and Change of Zooplankton in the Season of Spring and Autumn in Poyang Lake *Journal of Lake Sciences* 15(4) p 345-352

[9] Kevin Pauwels, Robby Stoks, Anne Verbiest, Luc De Meester 2007 Biochemical adaptation for dormancy in subitaneous and dormant eggs of Daphnia magna *Hydrobiologia* 594 p 91-96

[10] Maia-Barbosa PM, Eskinazi-Sant'Anna EM, Valadares CF, Pessoa GCD 2003 The resting eggs of zooplankton from a tropical, eutrophic reservoir (Pampulha Reservoir, south-east Brazil) *Lake & Reserviors: Research and Management*8p269-275

[11] Shen Jiarui, Fauna Sinica Arthropoda Crustacea Freshwater Copepoda 1979 BeiJing: Science Press, Academia Sinica, China

[12] Jiang Xiezhi, Du Nanshan, Fauna Sinica Crustacea Freshwater Cladocera 1979 BeiJing: Science Press, Academia Sinica, China

[13] Ruban V, López-Sánchez JF, Pardo P, Rauret G, Muntau H, Quevauviller Ph 2001Harmonized protocol and certified reference material for the determination of extractable contents of phosphorus in freshwater sediments – A synthesis of recent works *Fresenius J Anal Chem* 370 p 224-228

[14] Wang Zhengfang, Hu Chuanyi, Lu Haiyan 1999 A method for determine total nitrogen in sea water and marine sediments *Journal of Ocean University of Qingdao (Supplement)* p 161-166

[15] Environmental Protection Agency 1979 Methods for Chemical analysis of water and wastes. USEPA. Watshington, DC p 460

[16] Ter-Braak CJF and Verdonschot PFM 1995 Canonical correspondence analysis and related multivariate methods in aquatic ecology *Aquat Sci* 57 p 255

IOP Conf. Series: Earth and Environmental Science **51** (2017) 012024 doi:10.1088/1755-1315/51/1/012024

Effects of Irradiation on Microbial Community Structure in the Yangtze River and Selection of Representative Microorganisms

J B SONG[1] and L P KUAI[1]

1 School of Nuclear Science and Engineering, Shanghai Jiao Tong University, Shanghai 200240, China

243887682@qq.com , L P KUAI E-mail: sjtu-lkuai@qq.com;

Abstract. In the field of radioactive contamination, few studies were conducted on using microorganisms for biological monitoring. However, studies regard microorganisms as pollution indicator organisms are very common in the field of heavy metal and pesticide pollution. In this study, we chose the Yangtze River water as the research object, and studied the changes of species and quantity of cultivable bacteria in water samples under different irradiation by separation and purification methods. Results indicated that irradiation will cause the change of microbial community structure and the decline of their quantity; Among them, (flavobacterium) is mostly affected and (bacillus) is less affected. Finally, we found two representative microorganisms (Flavobacterium nitratireducens strain N1; Novosphingobium aquiterrae strain E-II-3) in the Yangtze Water and provided a reference for using microorganisms to monitor the radioactive pollution.

1. Introduction

At present, in order to ensure the required energy for economic development in China's inland region, the construction of inland nuclear power plants will become the inevitable demand [1]. Compared with coastal nuclear power plants, inland nuclear power plants' volume of the receiving water is less. Once the nuclear leakage accident happens, it will have great influence on local residents' production and life. Therefore, China has put forward strict requirements for inland nuclear power plants to achieve zero emission [2]. At the same time, inland nuclear power plants need to use more scientific monitoring standards to make a more objective evaluation [3]. Among them, biological monitoring can be used as an important means, which makes a more objective evaluation about the impact from radionuclides that are released from inland nuclear power plants under their normal operation, accident condition and retire process on the environment [4].

As the population which has the largest number, maximum biomass, and the biggest influence on the cycle of life elements in the nature, microbe has a unique advantage when used for monitoring water environment [5]. As the most basic member of the ecological system, microorganisms are more sensitive to pollution elements, such as heavy metals and organic matters in the environment. They can quickly feel a series of changes in the environment and make corresponding reactions through the related mechanism. Therefore, microorganism is considered as the most potential indicating organism [6].

Currently, in a series of fields polluted by heavy mental and pesticides, related researches choosing microorganisms as pollution indicating organisms have made remarkable achievements [7]. However, in the field of radioactive pollution, researching reports choosing microorganisms to do the biological monitoring are scarce. By comparing the changes of cultivable microorganisms before and after irradiation in the Yangtze River water sample, this paper researches the influence from irradiation on the structure of microbial population. Meanwhile, it looks for representative microorganisms in the water body and provides the corresponding reference for the biological monitoring of the radioactive pollution in Yangtze River basin.

2. Materials and Methods

2.1. Water Sample Collection

Referring to "Surface Water and Sewage Test Specification", we collected water samples in three points of Yangtze River area on April 15, 2016. The sampling location is shown in figure 1.

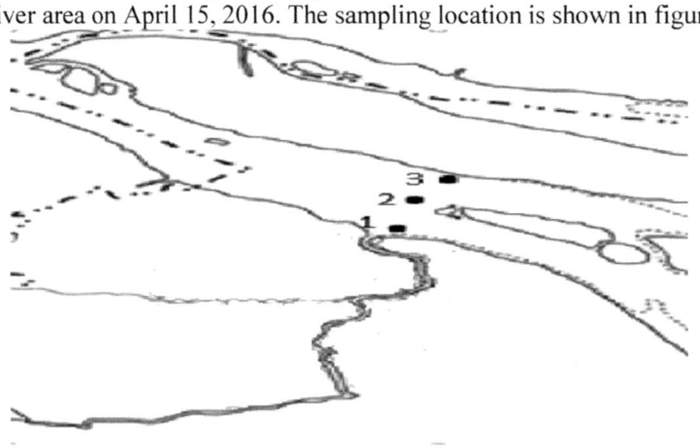

Figure.1 Location of sampling sites

We used organic glass sampler to collect the water sample under 0.5 m of the water surface, mixed the water sample, loaded it into plastic bottles disposed by moist heat sterilization in advance, put the sample in the heat preservation box, and immediately sent it to Shanghai Metrological Testing Technology Research Institute to do the irradiation.

2.2. Sample Irradiation

Sample Irradiation: Using 60Co to do the irradiation for the sample, with dose respectively 2 Gy、20 Gy、200 Gy and dose rate 4Gy/h, and taking a group sample without irradiation as the blank control.

2.3. Analysis Method

2.3.1. Counting bacteria in the sample. After being diluted to 10-1, 10-2, 10-3 and 10-4, the sample is coated on the agar plate and cultivated in 30℃ [8]. Everyday, we observe bacterial colonies' morphology and also do the community statistics. According to bacterial colonies' size, color, uplift situation and edge shape, we distinguish different types of strains.

2.3.2. *Single strain's physiological and biochemical function determination.* After transferring out the strain cultured from R2A table, we purify it three times. Referring to "Common Bacterial System Identification Manual" [9], we do the physiological and biochemical function experiment for single strain and preliminarily identify them.

2.3.3. *Single strain molecular biology identification.* Strain will be sent to Shanghai Personal Biotechnology Co., Ltd for DNA extraction, PCR amplification and sequence determination, and determine the strain.

2.3.4. *Correlation analysis.* Using software Origin Lab 9.0 to do the curve fitting about the total number of bacteria and the changing number of single strain under different irradiation doses, and understanding the correlation between the irradiation intensity and the changing number of the strain.

2.3.5. *Community diversity analysis.* Four community diversity indexes [10]: Community diversity abundance level (S), dominance level (P), Simpson index (L) and Shannon-Wiener index (H).

S: The number of different species of bacteria in the sample

P: The proportion of a certain type of bacteria,

$$p_i = n_i/N \tag{1}$$

L: Referring to the probability of the same type of bacteria selected from two colonies randomly. The smaller *L* indicates that the population distribution is more uniform and species are more abundant.

$$L = \sum_{i=1}^{S} [n_i(n_i - 1)/N(N - 1)] \tag{2}$$

H: The composite index that can reflect the community evenness and diversity. The larger *H* indicates that the biodiversity is higher.

$$H = -\sum_{i=1}^{S} \left(\frac{n_i}{N}\right) \ln(n_i/N) = -\sum_{i=1}^{S} p_i \ln p_i \tag{3}$$

3. Conclusion and analysis

3.1. Identification results

Eighteen kinds of bacterial colonies are separated from the water body and are named respectively as a-r. We can conduct physiological and biochemical experiments (Table 1) and molecular biological identification (Table 2) for all 18 kinds of bacterial colonies. The result shows that bacterial colonies from the separation belong to different species. Among them, the share of *Flavobacterium*, *Bacillus* and *Acinetobacter* is larger.

Table.1 Physiological and biochemical characteristics of strains

	a	b	c	d	e	f	g	h	i	j	k	l	m	n	o	p	q	r
Methyl test	-	-	-	-	-	-	-	-	-	-	-	+	-	-	+	-	-	-
VP test	-	-	-	-	-	-	-	-	+	-	-	-	-	+	+	-	-	-
Starch hydrolysis	-	-	-	-	+	-	+	-	+	+	-	+	+	-	+	-	-	+
Gelatin liquefaction	-	-	+	+	-	-	-	+	+	-	+	+	+	+	+	-	+	-
H_2S production	+	+	+	+	+	-	+	+	+	+	+	+	-	+	+	+	+	+
Catalase test	+	-	+	+	-	+	+	+	+	+	-	+	+	-	+	-	+	-
Nitrate reduction	-	+	+	+	-	-	-	+	+	-	-	+	-	+	-	-	+	+
Acid from Glucose(open)	+	+	-	+	-	-	+	-	-	-	+	-	+	-	+	-	+	-
Gas from Glucose(open)	-	+	-	-	-	-	+	-	-	-	+	-	+	-	+	-	+	-

Table.2 Bacterial identification results

Bacteria	DNA identification results	Identities
a	*Rhodococcus equi strain DSM 20307*	99%

b	*Novosphingobium soli strain CC-TPE-1*	98%
c	*Flavobacterium glycines strain NBRC 105008*	97%
d	*Flavobacterium nitratireducens strain N1*	98%
e	*Hymenobacter ocellatus strain Myx 2105*	93%
f	*Acinetobacter johnsonii strain ATCC 17909*	99%
g	*Bacillus stratosphericus strain 41KF2a*	100%
h	*Nocardioides marinus strain CL-DD14*	98%
i	*Bacillus aryabhattai strain B8W22*	98%
j	*Empedobacter halobium strain G393-B445*	99%
k	*Novosphingobium aquiterrae strain E-II-3*	97%
l	*Flavobacterium pectinovorum strain DSM 6368*	97%
m	*Porphyrobacter colymbi strain TPW-24*	99%
n	*Flavobacterium plurextorum strain 1126-1H-08*	98%
o	*Exiguobacterium acetylicum strain DSM 20416*	99%
p	*Acinetobacter johnsonii strain DSM 6963*	97%
q	*Brevibacillus agri strain NBRC 15538 16S*	99%
r	*Pelomonas saccharophila strain DSM 654*	98%

3.2. Irradiation's effects on microbial quantity

After the irradiation, we count the number of various forms of bacterial colonies on the plate and get the bacterial community composition as shown in table 3 under different irradiation intensities. You can see that with the increase of irradiation dose, the number of microorganisms in the sample has a declining tendency. The number of bacteria in the sample decreases from 2.4X104 CFU/ml in the control group to 6.8X103 CFU/ml under 200 Gy irradiation intensity, total decreased by 71.67%, which shows that high irradiation intensity will make the inactivation of most microorganisms. A Krisko's [11] experiments also come to the similar conclusion. However, when the irradiation dose is 2 Gy, the total number of bacteria and the number of bacteria in the control group are both around 2.2X104 CFU/ml. Both are flat, showing that the irradiation dose won't have extremely large effects on microorganisms.

Table.3 The flora composition of bacteria in different irradiance sample

Bacteria	a	b	c	d	e	f	g	h	i	j	k	l	m	n	o	p	q	r	Total
Dose	Number of colony-forming units(10^2CFU/ml)																		
200Gy	4	3	7	11	5	13	6	1	7	1	4	2	3	0	0	0	0	1	68
20Gy	1	6	15	42	12	35	2	2	12	0	9	0	3	2	3	1	4	0	149
2Gy	0	9	42	85	15	30	0	0	8	0	19	0	4	0	4	0	5	0	221
0Gy	0	6	34	95	8	22	0	5	8	0	36	0	10	0	5	0	6	0	235

3.3. Irradiation's effects on microbial species

According to the table 3, the analysis of cultivable bacterial communities' structure before and after the irradiation in the Yangtze River water shows that: When the irradiation dose is 2 Gy, the structure of bacteria community hasn't changed. It indicates that the irradiation of this dose has very small effects on microorganisms, which is consistent with the influence from irradiation on the number of bacteria. With the increase in irradiation intensity, the number of bacterial species goes up 4 instead. Since the irradiation leads to the death of lots of sensitive dominant strains, other strains get sufficient growth space. Table 4 shows that Simpson index (*L*) decreases with the increase of irradiation, which proves the conclusion that when the irradiation dose increase, the bacterial communities will become more diversified and the community distribution will become more even. WS Elsayed's [12] studies also have the similar situations.

In the Yangtze River water, the proportion of *flavobacterium* is very high (over 50%), belonging to a dominant strain. With the increase in the irradiation intensity, the proportion of the dominant strain, *flavobacterium*, declines from 59.29% to 29.31%, the total number decreased by 84.50% , which is affected greatly from the irradiation. By contrast, the proportion of *Bacillus* in the control group is 3.35% and goes up to 19.12% in the 200Gy group, the total number increased by 38.46%, showing the strong irradiation tolerance of this strain.

Table.4 The community diversity index in different irradiance samples

Irradiation dose	Colony richness(*S*)	Dominant bacteria	Percentage(*P*)	Simpson index(*L*)	Shannon-Wiener index(*H*)
200Gy	14	*Flavobacterium*	29.31%	0.109	2.391
20Gy	15	*Flavobacterium*	38.93%	0.166	2.913
2Gy	10	*Flavobacterium*	57.62%	0.219	2.114
0Gy	11	*Flavobacterium*	58.29%	0.223	2.179

3.4. Irradiation's effects on specific strains

After observing figure 2, we can find that the changing tendency of *flavobacterium* dominant strain d （*flavobacterium nitratireducens strain N1*） is very close to the changing tendency of the number of strains in the figure 3 . And with the increase of the irradiation dose, its proportion is 40.43% in the control group and declines to 16.18% in the 200 Gy group. What's more, 88.42% of strain d （*flavobacterium nitratireducens strain N1*） has disappeared. The negative correlation between the number of strains and the irradiation dose successfully reflects the changing situation of microbial quantity when the irradiation intensity changes. It also shows that we can select this strain to study as the representative microorganism in the Yangtze River water.

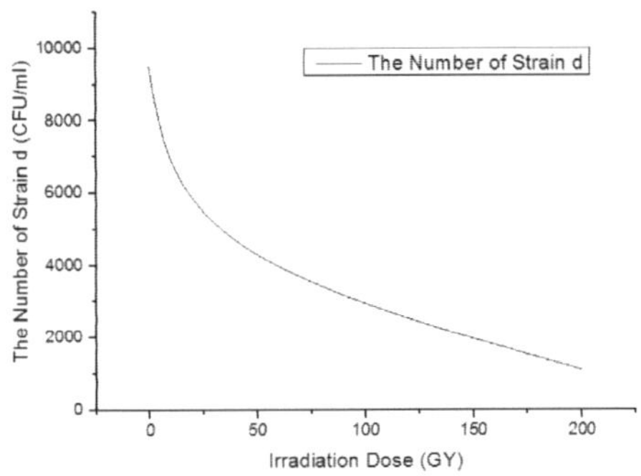

Figure.2 Variation of strain d under different irradiation doses

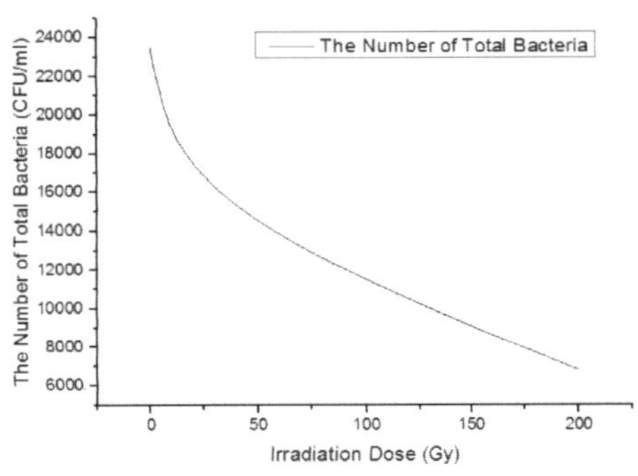

Figure.3 Variation of total bacteria under different irradiation doses

At the same time, as the Table 3 shows, the number of strain k（*novosphingobium aquiterrae strain E-II-3*）appears a declining tendency with the increase of the irradiation dose, which also can be study as a representative microorganism in the Yangtze River water.

4. Discussion

At present, most of domestic and overseas [13-14] studies about the influence of irradiation on microbes choose KGy dose. However, this experiment uses Gy magnitude to do the irradiation, where most microorganisms show a higher irradiation fatality rate. The main reason is that irradiation sample in this experiment is natural body, whose cultivable microbe concentration is only 2.35×10^4 CFU/ml. But other experiments choose the bacteria suspension experiencing the enrichment culture as the research object, whose cultivable microbe concentration is about 10^6—10^7 CFU/ml. Related researches [15] show that with a higher initial bacterium content, irradiation will have the smaller influence on microorganisms. In addition, this experiment chooses plastic containers. Compared with glass tubes used by M Barakatsm et al [14], plastic containers have a smaller density, whose atomic number is smaller and radiation permeability is better.

Damage for microorganisms from the irradiation has two types, one is the damage from materials, causing the disorder of enzyme system, physiological and biochemical delay or stop, and metabolic disorder or interrupt, and finally leading to the cell death. The second type of damage is from free radicals produced from the irradiation in the cytoplasm, leading to DNA single-stranded and double-stranded rupture and indirectly causing the death of microorganisms. Mainly through the rapid repair of damaged DNA fragments, microorganisms form spores to slow the metabolism rate and produce superoxide dismutase to remove free radicals this kind of ways to reduce the irradiation damage. *Bacillus* bacteria mentioned in this experiment can reduce moisture content through producing spores in adverse conditions and reduce metabolism rate to reduce irradiation damage, leading to a stronger irradiation tolerance. At the same time, Zhi Qixun [16] found that gram-positive bacteria have a higher radiation tolerance than gram-negative bacteria. Wujiang et al [17] came to the conclusion that the amount of superoxide dismutase in aerobic microorganism is higher than facultative anaerobic microorganism.

This paper selects d (*Flavobacterium nitratireducens strain N1*) and k (*Novosphingobium aquiterrae strain E-II-3*) which belong to facultative anaerobic non-spore gram-negative bacterium. Compared to other separated strains, they are more sensitive to the irradiation. Among them, strain d (*Flavobacterium nitratireducens strain N1*) belongs to the *Flavobacterium*. The result from Glockner F.O. et al [18] analysis of 16S rRNA in every country's water body shows that this kind of bacteria widely exists in the fresh water. Therefore, it has the researching value as the representative microorganism in the Yangtze River water.

Currently, related researches mainly focus on the influence from irradiation on the single strain in order to get irradiation resistant strain or find out the fatality rate caused by irradiation, such as LEENY et al [19] research about the irradiation lethal dose of Escherichia coli and bacillus cereus etc., and FA. Rainey et al [20] selection of irradiation resistant strain. This paper, however, focuses on the overall impact from irradiation on the cultivable microbes in the water body and explores the changing situation of microbe population's structure caused by irradiation. At the same time, this paper aims to look for the representative microorganism in the Yangtze River water and uses selected microorganism to provide the reference for radioactive pollution monitoring.

5. Conclusion

- Irradiation can lead to an increase in microbial community structure and the decline of its number in water body.
- Irradiation's influence on *Flavobacterium* is largest, but its influence on *Bacillus* is comparably smaller.
- Finding two kinds of representative microorganisms in the Yangtze river water, which are d (*Flavobacterium nitratireducens strain N1*) and k（*Novosphingobium aquiterrae strain E-II-*

3）, and choosing microorganisms to provide the reference for radioactive pollution monitoring in Yangtze River basin.

References

[1] Song M Q 2012 Analysis of the necessity of nuclear power development in inland China *Scientific and Technological Innovation and Application* **27** 178
[2] Guo X W, Tan B Q, Zhao J, 2013 Preliminary analysis of inland nuclear power for surface water quality impacts *Conservancy Technology and Economy* **19(4)** 5
[3] Beresford N A, Hosseini A, Brown J E 2010 Assessment of risk to wildlife from ionising radiation: can initial screening tiers be used with a high level of confidence *Journal of Radiological Protection* **30(2)** 265
[4] Copplestone D, Beresford N A, Brown J E 2013 An internationaldatabase of radionuclide concentration ratios for wildlife: development and uses *Journal of Environmental Radioactivity* **126(4)** 288
[5] Azarbad H, Niklińska M, CAM Van Gestel 2013 Microbial community structure and functioning along metal pollution gradients *Environmental Toxicology &Chemistry* **32(9)** 1992
[6] Ibekwe A M, Grieve C M 2004 Changes in developing plant microbial community structure as affected by contaminated water *Fems Microbiology Ecology* **48(2)** 239
[7] Zheng Y N, Wang S Y, Guo J H 2008 Trends in the use of protozoan and metazoan in the indication and assessment of WWTP operation *Microbiology* **35(12)** 1943
[8] Zhao B, Lin H, He S J 2014 *Microbiology Experiment (Beijing)* vol 2 (Science Press) p 58
[9] Dong X Z, Cai M Y 2001 *Systematic Determinative Manual of General Bacteria* (Beijing: Science Press) p 223
[10] Edwards L, Lilley A K, Timms-Wilson T H 2001 Characterisation of the culturable heterotrophic bacterial community in a small eutrophic lake *FEMS Microbiology Ecology* **35(3)** 295
[11] Krisko A, Radman M 2010 *Proceedings of the National Academy of Sciences of the United States of America* **107(32)** 14373
[12] Elsayed W S, Ghanem S 2009 Bacterial community structure change induced by gamma irradiation in hydrocarbon contaminated and uncontaminated soils revealed by PCR-denaturing gradient gel electrophoresis *Biotechnology* **8(1)** 78
[13] Qiu Y X, Ye S Y, Yu K 2007 Research on the lethal reason of radiation to Escherichia coli and Lactobacillus *Food Science and Technology* **6** 38
[14] Barakatsm M, Gary B, Richardh L 2010 Inactivation of Escherichia coli O157: H7, Listeria monocytogenes, Salmonella enterica and Shigella flexneri on Spinach Leaves by X-ray *Food Microbiology* **27(1)** 24
[15] Wei M, Yang P H, Li Q 2010 Main factors affecting sterilization effect of food irradiation *Science and Technology of Sichuan Agriculture* **3** 54
[16] Zhi Q Y 2010 *New Food Sterilization Technology* (Beijing: Agricultural Press) p 167
[17] Wu J, Chen D J 1997 Determination of SOD activity in microorganisms *Chinese Journal of Pharmaceuticals* **11** 495
[18] Frank Oliver Glöckner, E Zaichikov 2000 Comparative 16S rRNA Analysis of Lake Bacterioplankton Reveals Globally Distributed Phylogenetic Clusters Including an Abundant Group of Actinobacteria *Applied & Environmental Microbiology* **66(11)** 5053
[19] LEE N Y, JO C, SHIN D H 2006 Effect of γ-irradiation on Pathogens Inoculated into Ready-to-use Vegetables *Food Microbiology* **23(7)** 649
[20] Rainey F A, Ray K, Ferreira M 2005 Extensive diversity of Ionizing-Radiation-Resistant bacteria recovered from Sonoran desert soil and description of nine new species of the genus deinococcus obtained from a single soil Sample *Applied & Environmental Microbiology* **(71)9 5525**

Area Estimation and Distribution Analysis of Subsurface Flow Constructed Wetlands at Regional Scale --Take Guangzhou City for Example

S X Yuan[1,2,3], G L Tang [1], H X Xiong[1,2,3], J Chen[1], X L Yin[1] and G Q Huang[1,2,3]

[1] Guangdong Open Laboratory of Geospatial Information Technology and Application, Guangzhou Institute of Geography, 100 Xianlie Middle Rd, 510070 Yuexiu, Guangzhou, Guangdong, P. R. China; yuanshx@gmail.com

[2] Guangzhou Institute of Geochemistry, Graduate School of the Chinese Academy of Sciences, 510640 Wushan, Guangzhou, Guangdong, China; yuanshx@gmail.com

[3] University of Chinese Academy of Science, Beijing 100049, China;

hgq@gdas.ac.cn

Abstract. In this paper, Area of Constructed Wetlands (CWs) required for treatment of domestic sewage generated by 13 million people was calculated in accordance with the distribution of existing population in Guangzhou City and mathematical model of CWs. By comparing this with land use data, the distribution of constructed wetlands at construction regional scale was simulated with GIS. The results show that, Guangzhou generate about 3.88 million m^3 domestic sewage per day, which shall be treated with 59.37 km^2 CWs. Assuming that a single wetland bed is 300 m^2, total 197,905 wetland beds shall be required in the city. Based on the analysis and statistics on data of second national land survey of Guangzhou City with GIS, there are enough ponds, bare lands, other grasslands and other garden plots in Guangzhou that can be used for construction of regional scale CWs, but the distribution of available lands in different regions is uneven. Constructed wetlands at regional scale are mainly distributed around Baini Channel, Tianma River, Xinjie River, Liuxi River Valley, Zengjiang River Valley and on both sides of the Pearl River through Panyu and Nansha.

1. Introduction

Constructed wetland is a sewage treatment technology developed since the 1970s [1]. Sewage treatment with CW will provide better social, economic and ecological benefits. There are more than 600 CW projects in the U.S. for municipal, industrial and agricultural wastewater treatment [2]. At least 200 CW systems are in operation in Denmark, Germany and Britain [2], more than 80 CW systems also have been put into use in New Zealand [3]. The first research on CW sewage treatment in China was carried out in 1987 by Tianjin Institute of Environment Protection, which established the first reed wetland treatment system covering an area of 6 hectares and providing a treatment capacity of 1400 m^3/d[4]. After 20 years of research and extension, Sewage treatment with constructed wetland has developed a relatively mature theory and practice system. The natural sewage treatment technology uses biological-ecological method to overcome shortcomings of chemical sewage treatment method which can cause secondary pollution. Furthermore, CWs possess advantages such as low investment cost and construction and operational cost, high economic efficiency [5-6] and good sewage treatment effect [2, 7-10]. Meanwhile, CWs consist of landscapes such as water, plants, animals, land and channel, which conform people's aesthetic requirements and form wetland landscape for the city [11-14]. Therefore, constructed wetlands can produce great ecological benefits, and this enables it to become one of the important options for domestic sewage treatment in the process of ecological city construction.

Guangzhou is an economically developed city with large population. There are about 13 million permanent residents [15] generating about 1.41 billion m^3 [16] municipal domestic sewage per year. Treatment of large amount of domestic sewage and control of canal pollution is a serious problem that the government of Guangzhou has been confronting for years [17-20]. In recent years, the government of Guangzhou has actively implemented ecological city construction, many researches on relevant policies are making effort on the construction of "Flower City, Green City and Water City"[21-23]. The functions and characteristics of CWs fit with the general goal of constructing ecological Guangzhou. Assuming that all domestic sewage of Guangzhou is treated with CWs, then many science questions at intermediate level shall come up, e.g., "how much area of wetlands shall be required in order to achieve effective treatment based on existing population and the volume of generated sewage in Guangzhou?" and "how to locate and distribute the regional scale constructed wetlands in Guangzhou?".

Most CWs scholars focused on the various pollutants in wastewater, removal of pollutants, wetland plant selection, water conditions, and related topics, while urban planners want to know whether it's possible to put the CWs which have the effect of sewage purification and landscape characteristic into a megacity and how to do this. In this paper, by combining the existing CWs researches and GIS technology, we estimated the area of subsurface flow wetlands required for the treatment of existing sewage in Guangzhou based on practical situation, and analyzed the distribution of regional-scale constructed wetlands, so as to provide a reference for ecological infrastructure planning in the process of ecological city construction or relevant fields.

2. Experimental Section

Sewage treatment system by utilizing subsurface flow constructed wetlands is widely applied for sewage treatment [24]. Comparing to free water surface system, subsurface flow system has advantages such as lower land area required for the treatment of equal volume of sewage, simpler operation and maintenance[25], higher nitrogen removal efficiency[26-29] and better removal effect for pollution indexes such as Total Phosphorus, Biochemical Oxygen Demand (BOD), Chemical Oxygen Demand, Suspended solid and heavy metals[30-32]. Subsurface flow CWs mainly consist of three parts: substrate, plants and water distribution system [32]. Plants and microorganism play an important role in sewage purification of constructed wetlands, while temperature is an important key factor that influences the growth of wetland biology [33-35]. Under the premises above, this research proposes the subsurface flow wetland model as follows: domestic sewage first led to the grid pool for enhanced filtration and then led to the subsurface flow constructed wetland model with single wetland bed area $\leqq 300m^2$, length-width ratio<2 and bottom slope$<1\%$ for treatment[36]. Without regard to other influencing factors, the area of constructed wetlands required for the treatment of existing volume of the sewage in Guangzhou is calculated.

2.1. Method

2.1.1. Calculation formula for constructed wetland area. The research shows that, the surface area of subsurface flow constructed wetland can be calculated through BOD_5 [36-37], i.e., reducing BOD_5 value through constructed wetlands to allow BOD_5 to meet the effluent requirement of a certain region. It is deduced based on first-order biochemical reaction kinetics [38]. This paper utilizes the first-order biochemical reaction kinetics to deduce the surface area of subsurface flow constructed wetland with the formula below.

$$S = \frac{Q(lnC_0 - lnC_e)}{K_T H \alpha} \times 2 \quad (1)$$

In the formula, Q is sewage flow (m^3/d); C_0 is influent BOD concentration (mg/L); C_e is effluent BOD concentration (mg/L); H is aquifer depth (m), α is porosity of constructed wetland bed substrate, K_T is biochemical reaction rate constant of wetlands at temperature T, "2" is safety factor.

2.1.2. Calculation formula for domestic sewage flow. Domestic sewage flow in residential area is calculated [39] as follows:

$$Q = \frac{q \cdot N \cdot K_z}{1000} \quad (2)$$

In the formula, Q is sewage flow (m^3/d), q is average sewage volume per person per day [L/(person•d)], N is the population of a certain region, Kz is peaking variation factor.

2.1.3. Calculation formula for biochemical reaction rate constant of wetlands. Influence of temperature on reaction rate constant is calculated as follows [25]

$$K_T = K_{20} \times 1.01^{T-20} \quad (3)$$

In the formula, K_{20} is optimum biochemical reaction rate constant (d-1) at 20°C; T is the temperature of domestic sewage.

2.1.4. Regional scale constructed wetland distribution analysis. The location and distribution of constructed wetlands must be determined according to natural environment conditions such as geology, landform, hydrology and sewage outlet municipal planning, etc. [40] the terrain in Guangzhou is higher in northeast and lower in southwest. Types of landform include middle and lower mountains, hills, tablelands and plains. Lithology and structure have a great influence on the landform [41]. Located at the top of Pearl River Delta, Guangzhou is endowed with a developed water system. There are north main stream of Dongjiang, the front, back and west channel of Pearl River, and Humen, Jiaomen and Hongqimen (three passages to sea) around the city, and there are Liuxi River, Zengjiang and Pearl River Delta network channel and mountain stream through the city. There are 231 river branches with a total length of 913 km [42] in the area. This shows that the natural environment conditions in Guangzhou City are fairly ideal for the construction of constructed wetlands. However, Guangzhou as a megalopolis, its cities, designated towns and villages occupy most areas of the main districts. Intensive urban construction restricts the location of constructed wetlands.

This paper proposes to analyze the distribution of constructed wetlands in Guangzhou City at regional scale with GIS Software. Ponds, bare lands, grasslands and other garden plots are considered as available lands for wetland construction based on present situation of land use mainly. Distribution pattern is discussed through the comparative analysis of map spot dimension of available land and regional scale constructed wetland needs. Based on constructed wetland needs of various regions, available lands that have an area greater than 0.01 km^2 are screened with GIS, then distribution simulation is conducted on the condition that the available land for construction of constructed wetlands are 1 km to the nearest river system and soil sediment is non-sandy .

2.2. Parameters settings and data

2.2.1. CW area calculation formula parameters setting. BOD_5 concentration in domestic sewage of Guangzhou is 30-200mg/L [43]. According to relevant standard of Ministry of Environmental Protection of the People's Republic of China, influent quality shall not exceed 80 mg/L [44]. In this paper, only wetlands under highest sewage concentration (i.e., 200mg/L) condition are considered to conduct theoretical discussion. The effluent quality shall be that the content of BOD_5 in water after treatment is lower than 20mg/L[45]. Yu-Lan Peng and Jin-Yong Zhang studied the influence of water depth on plants and got the idea that optimum water depth for wetland bed is 0.5m [46-47], so the optimum water depth of 0.5m is selected as the water depth (H) applied in this paper. U.S.

Environmental Protection Agency suggests that, dense vegetation area of wetlands take porosity of 0.65-0.75 while wetlands with open water take porosity of 1.06[48]. Since urban wetland is characterized by relatively low water area, average vegetation biomass and relatively dense vegetation, the value of porosity α is 0.75.

2.2.2. Domestic sewage flow calculation formula parameters setting. Domestic sewage quota of residents shall be determined according to the water quota adopted at the place in question. Other things considered include the level of water supply and drainage facilities, in addition to the universal degree of drainage system inside the building, which is usually around 80%-90% of relevant water quota at the place in question. Guangzhou is a megalopolis. Guangdong Nom of Water Use (DB44/T1461- 2014) [49] states that the mean domestic sewage quota of residents in megalopolis is 200 L/d/person. The Guangzhou Economic and Social Development Statistics Bulletin 2014 indicate that, permanent residents at the end of the year are 13.08 million [15]. Value of peaking variation factor [50] of domestic sewage volume is shown in the table 1 below:

Table 1. Value of Peaking Variation Factor Kz of Domestic Sewage Volume. The mean domestic sewage quota of residents in Guangzhou is 200 L/d/person, the *Kz* is 1.5.

Average daily flow (L/s)	5	15	40	70	100	200	500	⩾1000
Kz	2.3	2.0	1.8	1.7	1.6	1.5	1.4	1.3

2.2.3. Biochemical reaction rate constant of wetlands calculation formula parameters setting. Research of Yu-Bo Cui et al. shows that, the biochemical reaction rate is constant at 20°C K_{20}=0.86d^{-1} [25]. Nan-Qi Ren et al. believes that optimum temperature for microbial growth is 25°C [33]. Annual mean temperature in Guangzhou in 2014wasabout 22.2°C. In the year, there are six months with a temperature higher than 25oC, the lowest temperature is 13.2°C and the highest temperature is 29.5°C [16]. This research assumes the lowest temperature as 13°C and the highest temperature as 30°C. The value of K_T at multiple temperatures is calculated.

2.2.4. Distribution simulation data. The simulation is mainly based on the land use data of Guangzhou which is obtained from the Second National Land Survey, and the soil type data which is from the Guangdong Digital Soil V2.0 operated by Guangdong Institute of Eco-environmental and Soil Sciences (GIESS).

3. Result

3.1. Area of constructed wetlands

3.1.1. Total area of constructed wetlands in the city. Formula (2) shows that, domestic sewage volume in Guangzhou City is 3.88 million m^3/d. Area of constructed wetlands required for treatment of domestic sewage generated by existing population is calculated by Formula (1) as shown in Table 2.

Table 2. Area of Constructed Wetlands Required for Treatment of Sewage Generated by Existing Population in Guangzhou

Water depth (m)	Porosity	C_0 (mg/L)	C_e (mg/L)	Temperature (°C)	Area (Km^2)
0.5	0.75	200	20	13	59.37
0.5	0.75	200	20	20	55.38
0.5	0.75	200	20	25	52.69
0.5	0.75	200	20	30	50.13

As seen in Table 2, when water depth is 0.50m, porosity is 0.75, influent BOD_5 is 200 mg/L, effluent BOD_5 is 20mg/L, temperature is 13°C, the area of subsurface flow CWs required for the treatment of sewage generated by existing population in Guangzhou is 59.37 km^2; when temperature reaches 25°C, the area of required CWs reduces to 52.69km^2; although when treatment temperature reaches 30°C, only 50.13km^2 CWs shall be required, but few months have a mean temperature higher than 30°C, so the value is just for reference. Calculated based on maximum treatment area, the population equivalent of CWs for sewage treatment in Guangzhou is about 4.59 m^2/person.

3.1.2. Area of constructed wetlands in each district. As seen in Table 3, the area of CWs required for the treatment of sewage generated by entire population in Guangzhou is 59.37 km^2, which is equivalent to the total area of Liwan (close to the size of Florida Everglades Nutrient Removal project), but only accounts for 0.80% of the total area of the administrative districts (including Zengcheng and Conghua) in Guangzhou. Area of CWs required in Yuexiu accounts for 15.50% of the area of the administrative district, and this proportion is the highest among all districts. Area of CWs required in Haizhu, Tianhe and Liwan respectively accounts for 8.04%, 7.08% and 6.91% of the area of respect administrative districts, which are also very high. Area of constructed wetlands required in Conghua accounts for only 0.14% of the area of the administrative district; these numbers is the lowest. Considering that area of single wetland bed doesn't exceed 300m^2, total 197,905 wetland beds shall be required in the city, and each wetland bed can be used for the treatment of the domestic sewage generated by 66 persons.

Table 3. Area of Constructed Wetlands Required For Treatment of Sewage Generated by the Population in Each District of Guangzhou City

District	Population (unit:10,000 persons)	Sewage flow (10,000m^3/d)	Area of required constructed wetlands (km^2)	Area of the administrative district (km^2)	Proportion in the administrative district	Constructed wetland bed (unit: piece)
Baiyun	226.57	67.97	10.41	795.79	1.31%	34687
Conghua	61.02	18.31	2.80	1974.50	0.14%	9342
Panyu	144.86	43.46	6.65	529.94	1.26%	22178
Haizhu	158.34	47.50	7.27	90.40	8.04%	24241
Huadu	96.48	28.94	4.43	970.04	0.46%	14771
Huangpu	46.67	14.00	2.14	90.95	2.36%	7145
Liwan	88.92	26.68	4.08	59.10	6.91%	13613
Luogang	39.61	11.88	1.82	393.22	0.46%	6064
Nansha	62.51	18.75	2.87	783.86	0.37%	9570
Tianhe	148.43	44.53	6.82	96.33	7.08%	22724
Yuexiu	114.09	34.23	5.24	33.80	15.50%	17467
Zengcheng	105.18	31.55	4.83	1616.47	0.30%	16103
Total	1292.68	387.80	59.37	7434.40	0.80%	197905

3.2. Regional scale constructed wetland distribution simulation

3.2.1. Conditions of lands for constructed wetlands in each district. Table 4 is obtained from analysis and statistics on data of second national land survey of Guangzhou with GIS. Part or whole of the ponds, bare lands, other grasslands and other garden plots can be used for the construction of CWs. By comparing the table with Table 3, we can see that there are enough lands in all districts for the construction of CWs except for Haizhu, Liwan, Tianhe and Yuexiu. A total of 7.27 km^2 CWs are required in Haizhu. At present, 8.69 hectares wetland parks have been constructed in the district. Construction needs can be met if part of the wetland parks is transferred to sewage treatment purpose and meanwhile Haizhu Lake and other land resources are utilized. Since the four types of lands above reach an area of 208 km^2 in Panyu and Nansha, domestic sewage generated by Liwan, Tianhe and Yuexiu can be transported and treated in these two districts, thus 16.14 km^2 CWs must be further constructed in Panyu and Nansha. Therefore, in view of the available land, there are sufficient conditions in Guangzhou City for the construction of enough CWs for domestic sewage treatment.

Table 4. Area and Proportion of Ponds, Bare Lands, Other Grasslands and Other Garden Plots and in Each District of Guangzhou City

Districts	Ponds (km^2)	Bare lands (km^2)	Other grasslands (km^2)	Other garden plots (km^2)	Total (km^2)	Area proportion of required wetlands
Baiyun	54.18	8.83	7.76	1.54	72.31	14.39%
Conghua	23.43	0.92	2.50	0.61	27.47	10.20%
Panyu	106.40	1.62	10.50	2.88	121.40	5.48%
Haizhu	0.54	0.06	0.39	0.00	0.99	737.56%
Huadu	106.91	11.71	9.00	21.66	149.27	2.97%
Huangpu	1.60	0.04	0.49	0.01	2.15	99.67%
Liwan	0.18	0.38	0.95	1.27	2.78	146.90%
Luogang	15.08	4.13	1.64	0.95	21.81	8.34%
Nansha	82.39	1.27	2.97	0.30	86.93	3.30%
Tianhe	1.87	0.89	1.00	0.12	3.87	175.95%
Yuexiu	0.02	0.00	0.00	0.00	0.02	30181.10%
Zengcheng	78.37	0.44	8.92	0.05	87.78	5.50%
Total	470.96	30.29	46.12	29.41	576.77	10.29%

[a] Data source: the second national land survey.

3.2.2. Distribution pattern of constructed wetland construction. According to 3.1.2, certain districts require very large area of constructed wetlands for treatment of their sewage. Total 10.41 km^2 CWs are required to be constructed in Baiyun District which has the largest population. Although this only accounts for 1.31% of the total area of the district, it is extremely difficult to implement planning for such a large area of wetlands. Analysis and statistics on more than 40,000 map spots of ponds, bare lands, other grasslands and other garden plots with GIS show that, there are few available lands with an area greater than 1 km^2(Figure 1). The area of most available lands is less than 0.1 km^2, therefore, regional scale CWs can't be constructed intensively but distributed dispersedly in the city.

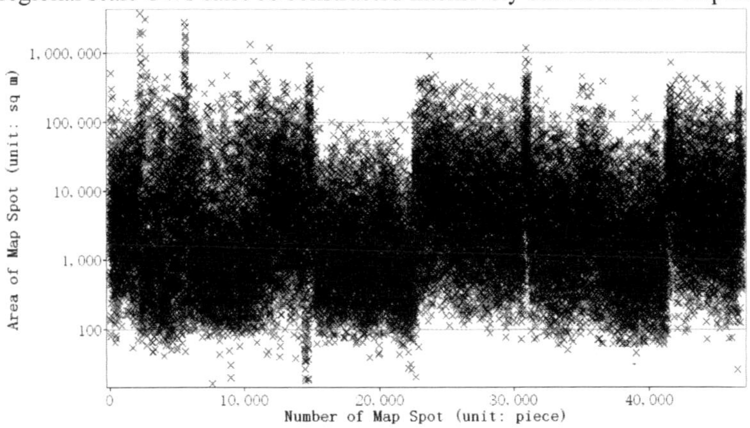

Figure 1. Scatter Diagram of Area of Available Lands. (Statistics on more than 40,000 map spots of ponds, bare lands, other grasslands and other garden, there are few available lands with an area greater than 1000000 m^2)

3.2.3. Regional scale constructed wetland distribution simulation. The results of simulation show that, main soil types in Guangzhou City include Ma Red Soil, Fluvial Alluvium Soil, Hydromorphic Paddy Soil, Ma Latosolic Red Soil etc. and Minor Sandy Soil. Most of the available lands are suitable for construction of CWs. CWs at regional scale are mainly distributed around Baini Channel, Tianma River, Xinjie River, Liuxi River Valley, Zengjiang River Valley and on both sides of rivers near Panyu and Nansha (Figure 2). There are enough lands in all districts except the five main districts - Haizhu, Huangpu, Liwan, Tianhe and Yuexiu for the construction of CWs. The total available lands in Huangpu are just enough for the construction of CWs, but insufficiency appears after inapplicable lands are excluded through analysis with GIS. Table 4 and Figure 2 show that there aren't enough CWs that can be constructed in the five main districts for the treatment of domestic sewage generated by their dense population. A few regional scale CWs can be distributed in the five main districts. Available lands in Panyu and Nansha shall be utilized to meet the most sewage treatment needs of the five main districts.

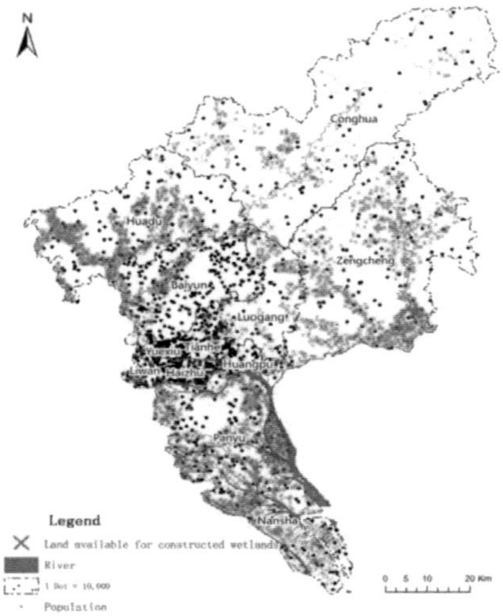

Figure 2. Distribution Simulation of Constructed Wetlands in Guangzhou at Regional scale (Land available for constructed wetlands: Including ponds, bare lands, other grasslands and other garden plots; Population: Population of Each District of Guangzhou City in 2014; 1Dot=10,000 Persons; "╳" This mark only shows the location of the available land)

4. Discussion

Compared to traditional sewage treatment plant, CWs possess advantages such as low investment and operational costs, high economic efficiency and good sewage treatment effect. Furthermore, CWs can serve as city wetland landscapes, increase ecological environment benefit [2, 5, 14] and play an important role in the improvement of water environmental quality of the city. On the basis of artificial lake, total 2.07 km^2 and 1.5 km^2 lake wetlands are newly constructed in Baiyun and Haizhu of Guangzhou respectively in recent years. Though construction of CWs is different from lake excavation, project quantities and engineering measures indicate that the construction of constructed wetlands is feasible to some extent.

The results show that population equivalent of CWs for sewage treatment in Guangzhou is about 4.59 m^2/person, which is close to the population equivalent 5 m^2/person specified in the "rule of thumb" for design of CWs [36, 51], this maybe shows that the calculated value of this article is close to the actual value. Upper limit of total area of CWs is not set in the Technical Specification of Constructed Wetlands for Wastewater Treatment Engineering [44], and the daily sewage treatment capacity of large-scale constructed wetland sewage treatment project is $\geqslant$10,000m^3/d, as the sewage volume generated in the districts of Guangzhou has gone far beyond this value. Large area of available lands shall be selected for wetland construction.

In this paper, area of subsurface flow CWs required for the treatment of domestic sewage generated by existing population of Guangzhou was calculated .The distribution of CWs for regional scale sewage treatment in Guangzhou is also briefly analysed with GIS under few limiting factors. However, since concentration of organic matters in sewage keeps changing [43], calculations conducted in this paper are based on single observed value as one cannot cover all possible situations. Many procedures such as pre-treatment, post-treatment, sludge treatment, odor gas treatment, auxiliary engineering facilities and supporting facilities [44] have to be considered in the practical construction of constructed wetlands. Therefore, the results of measurement and calculations made in this paper shall be provided just as a reference for ecological infrastructure planning in the process of ecological city construction or relevant fields.

5. Conclusions

Area of constructed wetlands required for the treatment of domestic sewage generated by existing population was calculated in accordance with domestic sewage flow in Guangzhou and mathematical model of existing CWs. The results show that, 13 million people generate about 3.878 million m3 domestic sewage per day, which shall be treated with 59.3715 km2 CWs. Assuming that a single wetland bed is 300 m2, total 197,905 wetland beds shall be required in the city. Analysis and statistics on data of second national land survey of Guangzhou with GIS show that, there are enough ponds, bare lands, grasslands and garden plots in Guangzhou that can be used for the construction of regional scale constructed wetlands, but the distribution of available lands in different regions is uneven. Most of the available lands are suitable for construction of CWs. Constructed wetlands at regional scale are mainly distributed around Baini Channel, Tianma River, Xinjie River, Liuxi River Valley, Zengjiang River Valley and on both sides of rivers near Panyu and Nansha. A few regional scale constructed wetlands are distributed in the five main districts. The insufficiency part can be distributed and constructed in Panyu and Nansha. Wetlands construction can be implemented in other districts selectively according to actual need of sewage treatment.

Acknowledgments

This work was supported by the National Natural Science Foundation of China under grant numbers 41271029 and 41301183, the Science and Technology Planning Project of Guangdong Province, China, under grant numbers 2013B060500067, 2013B060500068, 2013B060500072, and 2015B070701020, Special Funds of Research Platform Environment and Capacity Construction, Guangdong Academy of Sciences, under grant number 2016GDASPT-0103.

References

[1] Li Z M 2012 Research advances of constructed wetland *Journal of Green Science and Technology* **6** p 151-154
[2] Bai X H, Wang B Z, Yu M and Nie M S 1999 Development of constructed wetland wastewater treatment technology and its application in China *Journal of Harbin University of Civil Engineering and Architecture* **6** p 88-92
[3] Wu Y Y 2000 The application of artificial wet-land in New Zealand *Jiangsu Environment Science and Technology* **3** p 32-33.
[4] Peng C Y, Zhu G H, Yin G, Yu Y G and Zeng S S 2000 Research on constructed wetland wastewater treatment technology *Chongqing Environmental Science* **6** 43-45
[5] Drizo A, Frost C, Grace J and Smith K 1999 Physico-chemical Screening of Phosphate-removing Substrates for Use in Constructed wetlands Systems *Water Research* **33(17)** p 3595-3602
[6] Chi Y Z and Chen F L 2003 Practice of wastewater treatment of constructed wetland *China Water & Wastewater* **4** p 82-83
[7] Cai P Y and Ma X Q 2008 Study advance in wastewater treatment technology of constructed wetland *Subtropical Soil and Water Conservation* **1** p 8-11, 33
[8] Wu S B and Dong R J 2008 The application and research progress of constructed wetland for wastewater treatment *Water Treatment Technology* **8** p 5-9, 21
[9] Fan X H 2006 *Constructed wetland system for wastewater treatment and its application*

(Nanjing: Southeast University)
[10] Zhang L, Han Q and Sima W P 2007 A review of wastewater treatment by constructed wetland *Shanxi Architecture* **28** p 204-205
[11] Liu J H 2004 *Constructing the artificial wetland landscape in the urban green space* (Wuhan: Huazhong Agricultural University)
[12] Li J 2013 *Research on plant design in constructed wetland combining water purification in north china* (Beijing: Chinese Academy of Forestry)
[13] Yuan C F 2008 *Study on the constructed landscape mode of the artificial wetlands ecological system* (Chongqing: Chongqing University)
[14] Wu J 2012 *Urban constructed wetland landscape* (Nanjing: Nanjing Forestry University)
[15] Statistics Bureau of Guangzhou Municipality 2015 *Guangzhou Economic and Social Development Statistics Bulletin 2014* (Guangzhou) p 8
[16] Statistics Bureau of Guangzhou Municipality 2015 *Guangzhou Statistics Yearbook 2015* (Guangzhou: Statistics Bureau of Guangzhou Municipality) (2015/12/23) http://data.gzstats.gov.cn/gzStat1/chaxun/njsj.jsp
[17] Jin L H, Liang Z H , Wan Y L, Yuan J and Lan Y F 2005 Characteristics of urban canal water pollution and its control measurements *Urban Environment & Urban Ecology* **5** p 6-8
[18] Ye H P 2006 *Study on phosphorus contamination and rehabilitation of urban river surge* (Guangzhou: Graduate University of Chinese Academy of Sciences (Guangzhou Institute of Geochemistry, Chinese Academy of Sciences)
[19] Nanfang Daily 2008 *Why river odor in Yangcheng can't be cured over ten years* (Guangzhou: Nanfang Daily) (2008/11/19) http://news.sina.com.cn/c/2008-11-19/081214752460s.shtml
[20] Liu J S , Guo L C , Luo X L, Chen F R and Zeng Y P 2014 Characteristics of pollution chain of "rainwater-rainwater runoff-river" in Guangzhou urban districts *Environmental Chemistry* **6** p1040-1041
[21] Li S L 2015 Pattern of "flower city, green city and water city" is gradually formed in Yangcheng (NanfangDaily) (2015/03/24) http://epub.cnki.net/kns/detail/detail.aspx?FileName=NFRB20150324A0 21&DbName=CCND2015
[22] Xu H X, Sui F X and Sui J 2013 Present features of flower city, implement green city strategy and promote water city plan (Guangzhou Daily)(2013/10/25) http://epub.cnki.net/kns/detail/detail.aspx?FileName=GZRB201310250012&DbName=CCND2013
[23] Yue Shui Xuan 2014 Guangzhou City, Gangdong Province: Flower city, green city and water city (China Water Resources News)(2014/03/28) http://epub.cnki.net/kns/detail/detail.aspx?FileName=CSLB201403280062&DbName=CCND2014
[24] Zhang J Y, Xia S L, Xiong K and Jin X C 1998 Research on subsurface flow constructed wetland for wastewater treatment *Environmental Science* **4** p 38-41
[25] Cui Y B, Han X K and Song T H 2003 The Performance and design of constructed subsurface flow wetland for wastewater treatment *Environmental Science Trends* **2** p 23-25
[26] Huang J, Wang S H, Yan L, Liu Y and Wang F 2007 Research on Intensity of Nitrification and Denitrification in Subsurface-flow Constructed Wetlands *Environmental Science* **9** p 1965-1969
[27] Luo W G, Wang S H, Huang J and Qian W Y 2005 Research on Denitrification by Using Subsurface Constructed Wetland in Low Temperature *China Water & Wastewater* **8** p 37-40
[28] Luo W G 2005 *Research on nitrogen pollutants removal efficiency of subsurface flow constructed wetland* (Nanjing: Southeast University)
[29] Feng A K 2007 *Research on nitrogen removal function of subsurface flow constructed wetland* (Guangzhou: Guangdong University of Technology)
[30] Luo W G, Wang S H, Qian W Y, Huang J and Liu Y 2004 Study on the Phosphorus Removal Effect of Subsurface Constructed Wetland *Safety and Environmental Engineering* **4** p 21-25
[31] Song Z W, Guo B H, Han X Y and Cao J 2003 A Subsurface-Flow, Constructed Wetland for Wastewater Treatment and Application Thereof *Industrial Water & Wastewater* **34(6)** p 6-9

[32] Li S B, Shi P J and Guo L Z 2008 Anaerobic-Vertical Subsurface Constructed Wetland Used in Sewage Treatment *Technology of Water Treatment* **5** p 59-62
[33] Ren N Q, Ma F and Yang J X 2002 *Pollution Control Microbiology* (Harbin: Harbin Institute of Technology Press)
[34] Chen W Y, Yuan S X, Li H J, Lin F P, Li Q, Hu J M and Chen Z H 2010 Influences of temperature and light intensity on the growth and pollutant removal by Hydrilla verticillata *Acta Ecologica Sinica* **10** p 2639-2646
[35] Li W Y and Li Z Y 2012 Effect of Temperature and Substrate on Nitrogen and Phosphorus Removal by Constructed Wetlands *Water Conservancy Science and Technology and Economy* **11** p 1-4
[36] Vymazal J 2002 The Use of Sub-surface Constructed wetlands for Wastewater Treatment in the Czech Republic: 10 Years Experience *Ecological Engineering* **18(5)** p 633-646
[37] Reed S C 1993 Subsurface Flow Constructed wetlands for Wastewater Treatment: a Technology Assessment (Us Environmental Protection Agency, Office of Water) (1993/06/01) http://www.ceadu.org.uy/documentos/Manual%20EPA%20de%20Sistemas%20Naturales.pdf
[38] Xia H S, Cai M and Xiang X 2008 Study on Optimal Design of Constructed Wetlands *Yellow River* **7** p54-56
[39] Zhang Z H 2004 *Watersupply & Drainage Design Handbook: Urban Drainage* (Beijing: Beijing General Municipal Engineering Design & Research Institute)
[40] Chen K 2004 Study on Constructed Wetland Planning in NanChong City *Wetland Science* **4** p 314-319
[41] China Travel Geography Online 2013 Overview of Guangzhou landform and introduction of geological hazard easily happened (2013/01/29) http://news.cntgol.com/dlsy/20130129/12944.html
[42] Tianjin Water Resources Bureau 2016 Construction of river system in Guangzhou (2007/04/10) http://www.guangzhou.gov.cn/node_2382/node_503/node_658/2007-04/1176192726173583.shtml
[43] Hui Y Z, Li L and Shang G Y 2010 Research on Guangzhou Municipal Wastewater Quality Characteristics and Treatment Measures *Journal of Yunyang Teachers College* **3** p 23-25
[44] HJ 2005-2010 *Technical specification of constructed wetlands for wastewater treatment engineering* (Beijing: Ministry of Environmental Protection of PRC)
[45] DB4437-90 *Guangzhou City Sewage Discharge Standard of Local standard of Guangdong province* (Guangdong)
[46] Peng Y L, Tu W G, Bao W K, Gao X F, Wu N, Luo P and Xiao W Y 2008 Aboveground Biomass Allocation and Growth of Phragmites australis Ramets at Four Water Depths in the Jiuzhaigou Nature Reserve *Chinese Journal of Applied & Environmental Biology* **2** p 153-157
[47] Zhang J Y, Zhang J, Liu J, Liu D X, Wu H M and Xie H J 2012 Influence of water depth on running effect of surface-flow constructed wetland treating polluted river water *Chinese Journal of Environmental Engineering* **3** p 799-803
[48] Agency U.S.-Environmental-Protection 1990 *Free Water Surface Wetlands for Wastewater Treatment: A Technology Assessment* (Washingtong DC: BiblioGov) p 1-5
[49] DB44/T1461-2014 *Guangdong Nom of Water Use DB44T1461-2014* (Guangdong: Administration of Quality and Technology Supervision of Guangdong Province)
[50] GB50013-2014 *Code for design of outdoor water supply engineering* (Beijing: Ministry of Housing and Urban-Rural Development of the PRC)
[51] Vymazal J 2005 Constructed wetlands for wastewater treatment in Europe *Nutrient management in agricultural watersheds: a wetland solution. Wageningen, The Netherlands: Wageningen Academic Publishers* p 230-44

The Response of Lake to Glacier in Akesu River-Kaidu River Small Watershed during 30 Years

Hanwen Cui[1,2] and Qigang Jiang[3]

[1] Department of Computer Science and Technology, Zhuhai College of Jilin University, Zhuhai 519041, China
[2] Zhuhai Laboratory of Key Laboratory of Education Ministry's Symbol Computation and Knowledge Engineer, Zhuhai 519041, China
[3] College of geo-exploration science and technology, Jilin University, Changchun 130026, China

cuicui_0504@163.com

Abstract. Glacier is a sensitive indicator of the climate change, and also closely related to human beings. RS and GIS supply effective means to research the change of glacier. Those are important methods for compensating the shortage of previous ones. Small watershed is the basic unit of water circulation. Thus, the more reliable conclusion, which is the response of lake to glacier, will be obtained by the small watershed. In this paper, dividing the watershed into the smaller one, the Akesu River-Kaidu River small watershed was chosen as study area. Based on RS and GIS, under the climate change from cold-dry to warm-wet since 1977, 2000 and 2007, the variety of glacier was shown as decreasing firstly and then increasing a little. Meanwhile, the variety of lake was presented as increasing firstly and then decreasing greatly. However, the change tendency of glacier and lake was decreased. The decreased areas were 823.3 km^2 and 239.5km^2 separately. According to the spatial bivariate auto-correlation analysis map, the negative correlation between glacier and lake was more distinct. In Tianshan Mountain, owing to the change mainly influenced by elevation, glacier was decreased and lake was increased. In Bosten Lake, glacier was increased and lake was decreased due to the change mainly affected by climate.

1. Introduction

In the Western China, the topography and cold environment supply the condition to glacier and lake [1-2]. According to the previous research, under the global climate warming [3-5], the glacier is decreased sharply.

Glacier is the important part of mountain environment and water cycle [6-9]. It has the great effect on the water resource [10]. Watershed is the special natural region type, the water is considered as the core. It is also the basic unit of the water cycle. Taken watershed as the study unit can compensate the shortage of administrative division.

In the Western China, the first level watershed is too large to analyze the response of lake to the glacier. As a result, those were divided into small watershed, so that the more reliable research will be obtained.

In this paper, small watershed was chosen as study unit. Based on the RS and GIS, the dynamic variety of glacier and lake was researched. The response of lake to the glacier indicated by the spatial bivariate auto-correlation analysis.

2. Study area

In the watershed of the glacier distribution in the Western China, the Tarim River watershed is the largest one. Many high mountains located in this region. The Mt. Qiaogeli with the elevation of 8611m, which is the highest peak of the Karakorum Mountain, is also distributed. So the glacier in this region owns large amount and huge scale.

Among the source of Tarim River, the Akesu River is the largest. The small watershed, which is composed by Akesu River and Kaidu River was chosen as study area (Figure 1). The region was located in the northwest and north of the Tarim Basin, with temperate continental arid climate. The south section of Tianshan Mountain is distributed in west and north of the study area, the high peak is also in the region. Bosten Lake, which is the largest fresh water lake in China, is located in the east of the study area. Its main recharge source is Kaidu River.

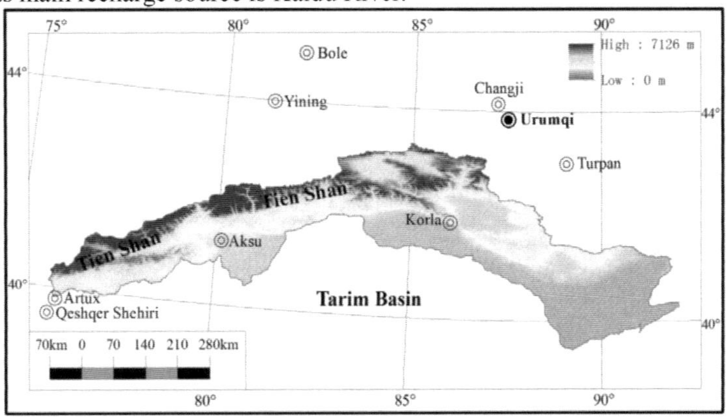

Figure 1. Study area

3. Method

According to the data about geology, geomorphology and hydrology, taking the Concise Glacier Inventory of China as reference, Western China was divided into 16 watersheds, and 50 small watersheds. Also, the ETM+ imageries and interactive interpretation technique were used.

This study uses MSS, ETM and CBERS-2 remote sensing imageries, with a time span of 30 years, since 1977, 2000 and 2007.

Firstly, images were preprocessed on the ERDAS software. In light of spectral characteristics and relative materials, combined with the previous studies, interpretation signs were set up. By using interactive interpretation technique, three periods' glacier and lake data were obtained in MapGIS platform.

Secondly, the glacier and lake's area change and spatiotemporal dynamic variety were approached by spatial analysis function of GIS so that the variety principle and distribution can be realized.

At last, a method of modified auto-correlation analysis was introduced in this paper, which was bivariate auto-correlation analysis method, used to investigate the spatial response of lake to the changes of the glacier. In terms of the LISA (Local Indicators of Spatial Association) method, which was put forward by Anselin [11], the local auto-correlation was chose to investigate the distribution of response relationships.

Local auto-correlation index formula is as follow:

$$I_p = \frac{(x_i - \overline{x}_i)}{\delta_i} \sum_j W_{ij} (x_j - \overline{x}_j) \quad (1)$$

x_i and x_j refer to observation value in location i and j separately. W_{ij} means to weight, and δ_i is the standard deviation of x_i. I's value is between -1and 1. The value greater than zero indicates positive correlation, less than zero indicates negative correlation, equal to zero indicates random distribution in space. The bigger the absolute value is, the higher the correlation would be.

4. Result

4.1. Area statistic

During the 30 years, from 1977 to 2007, the area of glacier decreased firstly in 2000, and then increased slightly. At the same time, lake increased firstly and then decreased. The decreased lake area in later period (from 2000 to 2007) was 3.2 times larger than the increased lake area in former period (from 1977 to 2000). Thus, during the research period, it had a tendency that the area of glacier and lake became decreased (Table 1).

Table 1. The statistic of glacier and lake area in study area (unit: km^2)

				area change	
object type	1977	2000	2007	1977-2000	2000-2007
glacier	5972.9	5064.2	5149.6	-908.7	85.4
lake	1282.6	1391.8	1043.1	109.2	-34.7

4.2. Spatiotemporal dynamic change

During the two research periods, according to the spatial distribution maps of glacier and lake's dynamic changes, glacier mainly located at high altitude, in the east and north of the study area, where the Tianshan Mountain is. Lake was mostly distributed at low altitude.

In the former period, decreased glacier area (947.6km^2) was as 24.3 times as large as increased one (39.0km^2). Significantly decreased glacier mainly located in the north of study area. And decreased glacier also distributed sporadically in other glacier location area (Figure 2 (1)).

In the later period, decreased glacier area (278.8km^2) was less than increased one (364.2km^2). Decreased glacier located sporadically along the glacier distribution area. Increased area mostly located in the east of study area (Figure 2 (2)).

Lake located in the study area, except the southeastern region. In the eastern area, where the glacier was large distributed, the lake was located in the low latitude.

In the former period, the decreased lake area (126.3km^2) represented 53.6% of the increased one (235.5km^2). Decreased lake mainly distributed in the southwest and north of the study area, and around the Bosten Lake. Increased area located in the north and south of the area, around the Bosten Lake, and the flat site of western region where is the foot of the Tianshan Mountain (Figure 2 (3)).

In the later period, decreased lake area (360.3km^2) was 31.3 times larger than increased one (11.5km^2). Decreased area distributed in the north and south of the area, and the flat site of western region where is the foot of the Tianshan Mountain. Around the Bosten Lake, the decreased area was larger than increased one obviously. Increased area mainly located in the north of the study area (Figure 2 (4)).

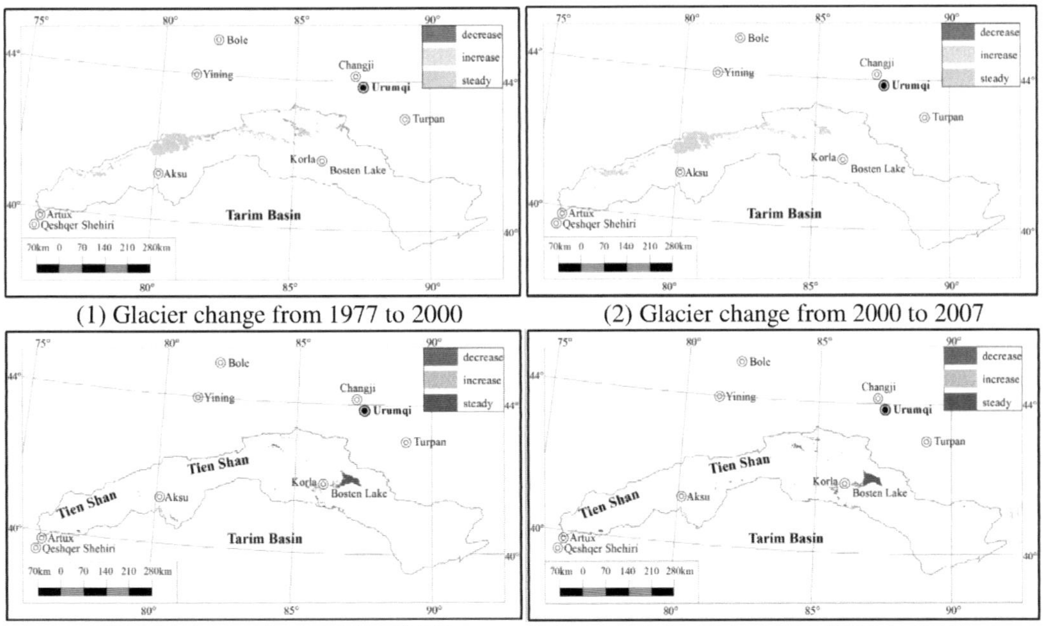

(1) Glacier change from 1977 to 2000 (2) Glacier change from 2000 to 2007

(3) Lake change from 1977 to 2000 (4) Lake change from 2000 to 2007

Figure 2. Spatial distribution maps of glacier and lake's dynamic changes

5. Response analysis

Precipitation, temperature and those combinations are the dominant factors influencing the dynamic variety of glacier and lake. The most important factor for glacier is the temperature. The second factor is the precipitation. Glacier's melting is decided by temperature, and accumulation is decided by precipitation. Making the comprehensive consideration, temperature, precipitation and elevation were considered to analyze the response of lake to glacier.

For the past 30 years, the annual mean temperature and precipitation had a curve upward tendency. They also showed the same rule in the 1977, 2000 and 2007 (Figure 3). The climate changed from cold-dry into warm-humid. It can explain the dynamic rule of the glacier well. Glacier mainly influenced by temperature in order that glacier gave priority to melt under the background of the rising temperature. However, with the increasing of the precipitation, glacier accumulated a little in the later period.

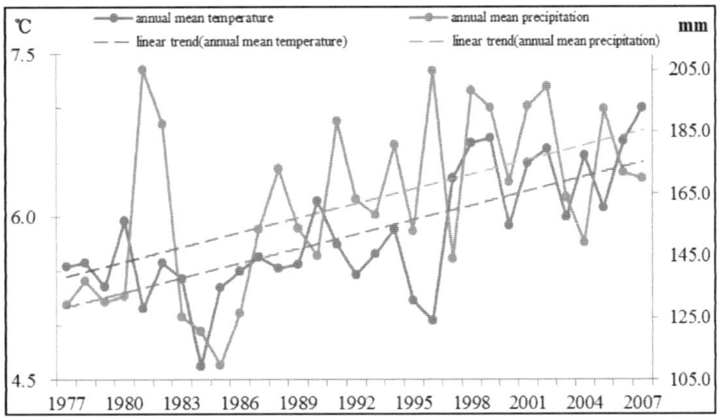

Figure 3. Annual mean temperature and precipitation change from 1977 to 2007

By means of the local index of the spatial auto-correlation analysis, the negative correlation between glacier and lake was more distinct. It was manifested as glacier decreased, lake increased (type one) and glacier increased, lake decreased (type two).

The response of increased lake to the decreased glacier located in the west and north of the study area, where the Tianshan Mountain located in (Figure 4). Those lakes were recharged mainly by melting glacier. Although the temperature was rising, but change of the lake was mainly influenced by elevation.

The response of decreased lake to increased glacier mostly distributed around Bosten Lake (Figure 4). In this region, the terrain is flat so that the correlation was mainly determined by climate factors. The Bosten Lake affected by its main recharge source, which is Kaidu River.

When the glacier melts to a certain degree, the ice sheet will thin out and the snowline will increase. Then, it will become a new balance. It is reduced that the amount of the small glacier at low altitude, which is most sensitive to the temperature change. In addition to slightly increased glacier, the recharge effect of melting glacier for Kaidu River is weak than before. As the Kaidu River's reduction, the Bosten Lake will be decreased. Previous research showed that the flow of the Kaidu River was reduced in 2002. That verifies the reason analysis of the response relationship in this paper.

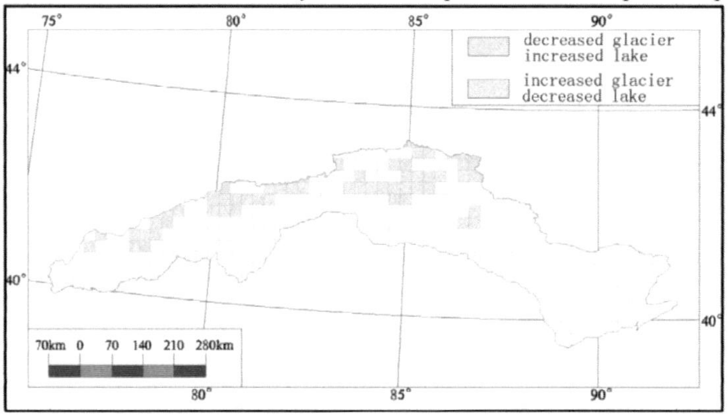

Figure 4. The response of lake to glacier map from 1977 to 2007

6. Conclusions

(1) During the 30 years, the variety of glacier was shown as decreasing firstly and then increasing a little. Meanwhile, the variety of lake was presented as increasing firstly and then decreasing greatly. However, the change tendency of glacier and lake was decreased. The decreased areas were 823.3 km^2 and 239.5km^2 separately.

(2) Glacier was mainly distributed in the north and west of the study area, where the Tianshan Mountain located. Lake was distributed in the low altitude of the small watershed beside of the southeast area.

Glacier was decreased significantly in the former period, and then increased remarkably in the later period. Those were located along the glacier distribution area and in the west of the small watershed separately. Lake was increased significantly in the former period and then decreased remarkably in the later period. Those were located in the flat site of western region, where is the foot of the Tianshan Mountain, and around the Bosten Lake.

(3) Annual mean temperature and precipitation had a curve upward tendency. Glacier was influenced by temperature more than precipitation.

(4) According to the spatial bivariate auto-correlation analysis map, the negative correlation between glacier and lake was more distinct. In Tianshan Mountain, owing to the change mainly influenced by elevation, glacier was decreased and lake was increased. In Bosten Lake, glacier was increased and lake was decreased due to the change mainly affected by climate.

References

[1] Wang X D, Zhong X H, Liu S Z 2008 *Journal of Arid Environments* **72** 1929-1939
[2] Jansson P, Hock R, Schneider T 2003 *Journal of Hydrology* **282** 116-129
[3] Houghton J T, Ding Y, Griggs D J 2001 *Climate change 2001: the scientific basis* (Cambridge: Cambridge University Press)
[4] Hodge S, Trabant D, Krimmel R 1998 *Journal of Climate* **11** 2161-2179
[5] Kang S C, Xu Y W, You Q L 2010 *Environmental Research Letters* **5** 1-9
[6] Owen A L, Thackray G, Anderson S R 2009 *Geomorphology* **103** 158-171
[7] Kargel J S, Abrams M J, BISHOP M P 2005 *Remote Sensing of Environment* **99** 197-219
[8] Paul F, Kääb A, Maisch M 2002 *Annals of Glaciology* **34** 355-361
[9] Qin D, Xiao C 2009 *The European Physical Journal Conferences* **1** 19-28
[10] Bajracharya R S, Mool K P, Shrestha R B 2007 *Impact of climate change on Himalayan glaciers and glacial lakes case studies on GLOF and associated hazards in Nepal and Bhutan* (Kathmandu: International Centre for Integrated Mountain Development G.P.O. Box)
[11] Anselin L 1995 *Geographical Analysis* **27** 93-115

The Hydrological Regimes Brought by the Three Gorges Project Affected Riparian Vegetation Distribution and Diversity in 2009 and 2010

Ling-Feng Miao [1,#], Wei-Wei Liu[2,#] and Fan Yang [1,*]
1, College of Environment and Plant Protection, Hainan University, Haikou, Hainan Province, China, 570228;
2, Lijiang Forest Ecosystem Research Station, Kunming Institute of Botany, Chinese Academy of Sciences, Lijiang, Yunan 674100, P. R. China
#, equally contributed to this work

*E-mail: yangfan@hainu.edu.cn; fanyangmlf6303@163.com

Abstract Post-dam riparian vegetations affected by the new hydrological regimes in the Three Gorges Reservoir (TGR) were investigated in 2009 and 2010, respectively. The investigation in 2009 showed that about 231 vascular plant species belonging to 169 genera of 61 families were distributed in the water-level-fluctuation zone (WLFZ) of the (TGR). Three vegetation types, including Chuanjiang, Gorge, and other vegetation types, were classified efficiently via cluster analysis. Alpha diversity analysis indicated that species richness gradually decreased with decreasing elevation. Beta diversity analysis indicated that high environment heterogeneity was existed between the lower section and the other two sections, and environment homogeneity was also existed between middle section and upper section. Using the analysis of the field growth in the 2009 and 2010 field surveys as bases, we proposed a list of perennial herb species and woody species that may potentially occurred in the WLFZ of the TGR. In addition, we predicted plant community structural changes in the different altitude sections of WLFZ in the future.

1. Introduction

The Three Gorges Project (TGP), as the largest hydropower project in the world, was initiated in 1994. The closure of the Three Gorges Reservoir (TGR) and the first impoundment was occurred in 2003. The water level of the reservoir rose to 135 m from about 60 m of the original riverbank in June

2003[1]. The second impoundment was impounded in October 2006, and the water level rose to 156 m. The third impoundment in October 2008 resulted in a sustained water level at above 170 m for five months. Normally, the natural peak flows in the river occur during July, August, and September (summer flooding) with low flows in January, February, and March [2]. However, the full functional Reservoir would have a water level fluctuating between 145 m and 175 m, i.e., 145 m in summer (May–October) for flood control and 175 m in winter (November–April) for energy generation [3, 4]. The analysis of hydrological regimes of the TGR is illustrated in another paper [4]. Thus, the new hydrological regime brought about by the TGP is the absolute opposite of the natural flood rhythms of the Yangtze River.

Although the TGP offers many benefits such as controlling flooding, generating motive power and electricity, facilitating navigation, and stimulating the development of local tourism, it simultaneously imposes negative and irreversible effects on the ecology and environment, especially on riparian ecosystems [3, 5]. The new hydrological regime from the TGP dramatically altered the conditions in the riparian ecosystems. The area between the high (175 m) and low (145 m) water level is called the water-level-fluctuation-zone (WLFZ)——a new kind of wetland[3]. According impounded conditions and the seasonal hydrological regimes in TGR [4], WLFZ was divided three sections. Upper section is an area between altitudes of 165 m-175 m, the middle section is an area between altitudes of 156 m-165 m, and the lower section is an area between altitudes of 145 m-156 m. Therefore, the riparian vegetation located within the WLFZ of the TGR suffered from different gradient flood stress. Of particular concern to us is determining the effect of the TGP on the riparian vegetation through new hydrological regimes. Therefore, the objectives of this paper are to (1) understand plant species composition and dynamic distribution of riparian vegetation in the WLFZ of the TGR; (2) determine the plant diversity in the different altitude sections of WLFZ; (3) provide useful information on selecting plant species with tolerance to winter flooding, and suggestions for revegetation in the WLFZ of the TGR; (4) predict plant community structural changes in the different altitude sections of WLFZ in the future.

2. Methods

2.1 Study area

The portion of the Yangtze River Basin between Chongqing and Yichang City (where the Three Gorges Dam is located at Sandouping) has a distance of approximately 600 km. The total area of the Three Gorges Reservoir Area (TGRA, 106°20′–111°50′ E, 29°16′–31°25′ N) is approximately 58,000 km^2 [5], and the total surface area of the rreservoir is 1,045 km^2. When water level was artificially regulated between 145 and 175 m, the WLFZ with an area of 348.93 km^2 was formed in the TGRA. The TGRA has a humid subtropical monsoon climate, with a mean annual temperature of 15–19°C, mean annual precipitation of 1,250 mm, and relative humidity of 76% [5]. Although the climate of the region is considered to be monsoonal subtropical, the flora is characterized by a mix of tropical, subtropical and temperate species [2]. Little affected by the quaternary glaciations, the TGRA is one of the richest areas in biodiversity in China [3, 6].

2.2 Field surveys

A survey on plant species distribution was carried out between July and August 2009. 22 sampling sites were set up in 10 typical areas along both sides of the Yangtze River in Zigui, Badong, Wushan, Fengjie, Yunyang, Wanzhou, Zhongxian, Fengdu, Fuling, and Changshou Districts of Chongqing City. The spatial distributions of these counties in the TGRA were shown in Yang et al. (2015). Three sampling belts (200 m in length, and 5 m in width) in each sampling sites were designed in the lower section, middle section and upper section of WLFZ, respectively, total 66 sampling belts. The plant species catalogue was recorded in detail.

A survey on plant community was carried out between June and July 2010. Sixteen sampling sites were listed in table 1. Three big quadrats (10 m × 10 m) were designed in the lower section, middle section and upper section in each sampling site, respectively. In addition, four small quadrats (1 m × 1 m) for herb species were designed in a big quadrat along the diagonal line. A total of 192 quadrats were used (Table 1). Basic parameters, such as altitude, slope degree, and vegetation coverage of each unit were recorded. For herbs and shrubs, species, abundance, coverage, and average height were investigated.

Table 1. The coordinate data of 192 quadrats of the water-level-fluctuation zone in the Three Gorges Reservoir in 2010

No. of plots	Site	No. of quadrats	Longitude (E)	Latitude (N)	Elevation	No. of plots	Site	No. of quadrats	Longitude (E)	Latitude (N)	Elevation
		1~4	106°26'51.0"	29°21'24.3"	153			97~100	108°47'12.3"	30°55'38.3"	148
1	Banan	5~8	106°26'53.0"	29°21'28.7"	162	9	Yunyang	101~104	108°47'14.9"	30°55'33.6"	160
		9~12	106°26'54.3"	29°21'31.3"	168			105~108	108°47'17.2"	30°55'29.7"	170
		13~16	107°05'15.0"	29°47'00.8"	154			109~112	109°24'13.2"	31°00'49.8"	152
2	Changshou	17~20	107°05'14.8"	29°46'53.9"	160	10	Fengjie 1	113~116	109°24'13.0"	31°00'53.6"	160
		21~24	107°05'15.1"	29°46'50.4"	170			117~120	109°24'13.1"	31°00'57.3"	170
		25~28	107°19'49.0"	29°44'27.6"	150			121~124	109°34'33.6"	31°02'45.4"	150
3	Fuling	29~32	107°19'48.9"	29°44'29.9"	162	11	Fengjie 2	125~128	109°34'36.7"	31°02'45.3"	162
		33~36	107°19'49.0"	29°44'33.7"	170			129~132	109°34'38.9"	31°02'45.1"	170
		37~40	107°50'05.9"	29°59'05.9"	148			133~136	109°54'51.2"	31°03'32.0"	148
4	Fengdu	41~44	107°50'07.0"	29°59'00.7"	162	12	Wushan	137~140	109°54'51.5"	31°03'28.4"	158
		45~48	107°50'10.3"	29°58'58.4"	170			141~144	109°54'51.3"	31°03'25.8"	170
		49~52	108°09'12.9"	30°23'51.1"	145			145~148	110°18'03.6"	31°01'58.7"	148
5	Zhongxian 1	53~56	108°09'15.1"	30°23'47.6"	164	13	Badong	149~152	110°18'07.4"	31°01'54.3"	160
		57~60	108°09'17.7"	30°23'44.3"	170			153~156	110°18'10.2"	31°01'51.9"	172
		61~64	108°05'22.6"	30°22'28.4"	148			157~160	110°45'19.9"	31°12'05.2"	148
6	Zhongxian 2	65~68	108°05'22.9"	30°22'24.7"	158	14	Xingshan	161~164	110°45'22.7"	31°12'05.2"	160
		69~72	108°05'22.7"	30°22'19.9"	170			165~168	110°45'24.9"	31°12'05.1"	170
		73~76	108°25'40.7"	30°43'34.8"	148			169~172	110°40'47.6"	31°03'50.6"	150
7	Wanzhou	77~80	108°25'46.1"	30°43'34.9"	156	15	Zigui 1	173~176	110°40'48.1"	31°03'50.9"	160
		81~84	108°25'49.3"	30°43'34.5"	165			177~180	110°40'48.4"	31°03'51.1"	172
		85~88	108°28'42.8"	31°09'00.4"	155			181~184	110°55'12.6"	30°51'55.4"	148
8	Kaixian	89~92	108°28'39.2"	31°09'00.1"	165	16	Zigui 2	185~188	110°55'11.7"	30°51'55.3"	160
		93~96	108°28'37.6"	31°09'00.9"	172			189~192	110°55'11.2"	30°51'55.4"	170

2.3 Hierarchical cluster analysis of plant species from different sampling sites.

Hierarchical cluster analysis was used to compare the similarity of plant species composition among different sampling regions. The plant species in each sampling site were scored as present (1) or absent (0). The coefficients of inter-sampling sites were generated. Hierarchical cluster analysis was performed with the SPSS 16.0 for Windows statistical software package.

2.4 Plant diversity index

Relative frequency = frequencies of a plant species/ frequencies of all plant species×100%;

Relative height = heights of a plant species/total heights of all plant species×100%;

Relative coverage = coverage of a plant species/ total coverage of all plant species×100%;

Important value = (relative frequency+ relative height+ relative coverage)/3.

The α-diversity parameters were calculated for each plot, and the average value of the plots at each site was accepted as the diversity indices of each site. The Shannon information index was used to measure diversity on the basis of importance value, the Pielou's index for the relative species evenness and the Simpson's index for the concentration of species dominance were also calculated, and richness was calculated as the species number. Different diversity indexes were calculated using the following equations [7, 8]:

Richness index (S): the amount of species in the plot;

Shannon-Wiener diversity index: $H' = -\sum_{i=1}^{s} Pi \ln Pi$;

Simpson's diversity index: $D = 1 - \sum_{i=1}^{s} Pi^{2}$;

Pielou evenness index: $Jsw = (-\sum_{i=1}^{s} Pi \ln Pi) / \ln S$;

Population encounter incidence: $PIE = \sum_{i=1}^{s} [(Ni / N)(N - Ni)/(N - 1)]$.

Where *Pi* is the relative importance values of the *i*th species, *Ni* is an importance of the *i*th species; *N* is total amount of importance values of each species in *i*th species present squadrat, and *S* is the total amount of the species in each squadrat.

The β-diversity for each section was calculated on the basis of data from all plots. The β-diversity values thus reflect habitat heterogeneity between sections. Different diversity indexes were calculated using the following equations [7, 8]:

Whittaker index: $\beta_W = S/\alpha - 1$;

Cody index: $\beta_C = [g(H)+l(H)]/2$;

Routledge index: $\beta_R = [S^2/(2r+S)] - 1$;

Where S is total amount of plant species in studied system, α is mean numbers of species in individual squadrat, g(H) is increased species number along the environmental gradient, l(H) is lost species numbers along the environmental gradient, and r is log of repeated species in each squadrat.

2.5 Statistical analysis

One-way analysis of variance (ANOVA) was used to test differences ($P<0.05$). Statistical analyses using the Duncan's method were performed with the SPSS16.0 for Windows statistical software package.

3. Results

3.1 Plant composition

The riparian vegetation survey in 2009 indicates that 231 vascular plant species belonging to 169 genera of 61 families were found in the investigated area. As far as family was concerned, family with over 30 species was not found. 29 families with one species and 26 families with 2-9 species were found, respectively, and their individual ratio to all families is 47.5% and 42.6%, respectively. As far as genus was concerned, genus with over 10 species was not found. 134 genera with one species and 33 genera with 2-5 species were found, respectively, and their individual ratio to all genera was 79.2% and 19.5%, respectively. As far as life form of plant species was concerned, 105 annual herb species and 75 perennial herb species was found, respectively, and their individual ratio to total species was 45.5% and 32.5%, respectively. In addition, 39 woody plant species including 12 tree species and 27 shrub or subshrub species were found.

3.2 Cluster analysis of vegetation types on various sites

The hierarchical cluster analysis indicated that the coefficient of vegetation types among different sampling sites. Figure 1 shows the results of the analysis. The results infer that vegetation types among different sampling sites may be differentiated. Thus, three vegetation types were differentiated efficiently. Type 1 was located in the Chuanjiang River sections (sites 1, 2, 3, 4, 5, 6, 9, and 10; from Chongqing to Fengjie); type 2 was located in the Gorges River sections (sites 14, 15, 16, 18, and 20; from Fengjie to Zigui); type 3 was located the experimental base (sites 22, 7, and 21) or other micro-topography (8, 11, 12, 13, 17, and 19) belonging to untraversed sites.

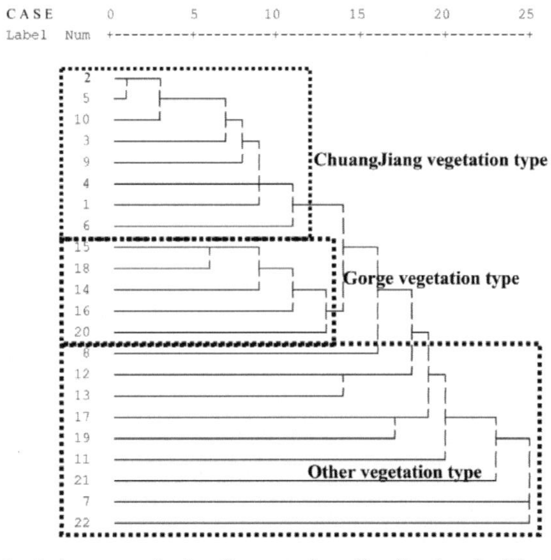

Figure1. Hierarchical cluster analysis of vegetation distribution in 22 sampling sites in the water-level-fluctuation-zone in the Three Gorges Reservoir in 2009.

Note: 1, Longxi River in Changshou; 2, Changming Port in Changshou; 3, Wu River in Fuling; 4, Yangtze River in Fuling; 5, Yangtze River in Fengdu; 6, Long River in Fengdu; 7, Experiment base in Zhongxian; 8, Xintian Town in Wanzhou; 9, Wanzhou Port in Wanzhou; 10, Pengxi River in Yunyang; 11, Yangtze River in Yunyang; 12, Yangtze River in the old location of Yunyang; 13, Yangxi River in the old location of Yunyang; 14, Meixi River in Fengjie; 15, Yangtze River in Fengjie; 16, Daling River in Wushan; 17, Yangtze River in the Badong; 18, Badong Port in Badong County; 19, Shennong River in Badong; 20, Xiangxi River in Zigui; 21, Lanling River Experiment base in Zigui; 22, Experiment base in Three Gorges Dam in Zigui.

3.3 Alpha diversity analysis among different altitude sections

The mean species richness based on the squadrats in 2010 was listed in Table 2. From upper section to lower section, species richness and diversity was gradually decreased with the decline of altitude. The best and the worst species richness were found in the upper section and lower section of WLFZ, respectively. Significant differences in species richness were found between middle section and lower section. However, significant differences in species richness were found between upper section and middle section. As far as diversity indexes were concerned, H′ and D diversity indexes in the lower section were obviously lower than that in the middle and upper sections. However, significant difference in H′ and D diversity indexes were not found between middle section and upper section. As far as evenness indexes were concerned, Jsw evenness index in the upper section was significant higher than that in the lower section, and Jsw evenness index in the middle section was not significant different from these in the upper section and lower section. As far as PIE was concerned, PIE in the lower section was significant higher than these in the middle and upper sections, and significant differences were not found between the middle and upper sections.

Table 2. The comparison of α diversity index in the different sections of the water-level-fluctuation zone in the Three Gorges Reservoir. Values are means ± SE. Within a column, values followed by different letters are significantly different at $P < 0.05$ according to Duncan's test.

	S	H′	D	Jsw	PIE
Upper section	16.9±6.3a	2.574±0.349a	0.907±0.031a	0.932±0.016a	1.465±0.192b
Middle section	14.1±5.6a	2.386±0.415a	0.885±0.054a	0.928±0.033ab	1.532±0.322b
Lower section	5.5±5.0b	1.312±0.643b	0.647±0.180b	0.898±0.057b	2.467±1.211a

3.4 Beta diversity analysis among different altitude sections

As far as vertical direction, β diversity in the lower section are significant different from upper section and lower section (Table 3). β_W, β_C and β_R in the upper section and middle section were significant higher than that in the lower section, and significant differences in these indexes were not found between upper section and middle section.

Table 3. The comparison of β diversity index in the different sections of the water-level-fluctuation zone in the Three Gorges Reservoir. Values are means ± SE. Within a column, values followed by different letters are significantly different at $P < 0.05$ according to Duncan's test.

	β_W	β_C	β_R
Upper section	2.373±0.336a	11.800±3.516a	19.657±6.175a
Middle section	2.465±0.411a	9.633±2.705a	16.340±4.945a
Lower section	1.562±0.595b	3.400±2.888b	5.685±5.367b

4. Discussion

Erosion and sedimentation processes are strongly linked to the hydrology of a river and also holds consequences for the surrounding vegetation. The riparian vegetation functions as filters between land and water and plays important roles in stabilizing riverbanks, which retains soil and decreases silt run off into rivers, and improving water quality [9, 10]. The riparian vegetation can be structured and to an extent, determined, by flood timing, duration, frequency, rate of change and magnitude, as plants often display a preference or tolerance for a specific hydrological regimes [2, 11]. The riparian ecosystems and plant diversity of the TGRA have been altered by the new hydrological regime caused by the TGP. The most serious effects on riparian vegetation are primarily caused by the opposition to the natural flood rhythms, increased flooding durations, and water level fluctuations with a 30 m magnitude. The riparian vegetation in the TGRA must not only endure winter flooding but also adapt to summer drought conditions during the growing season. Therefore, the analysis of vegetation responses to river regulation through TGP is highly important to assessing the effects on riparian ecosystems. Additionally, the data can be used as a starting point for monitoring purposes because it can be assumed that changes in vegetation will continue to occur in the future.

The investigation of pre-dam riparian vegetation in 2001 manifested that about 405 vascular plant species, belonging to 240 genera of 83 families, located in the TGRA [12]. The comparative analysis of the vascular plant family, genera, and species identified that a large proportion of the pre-dam riparian vegetation disappeared. Compared with pre-dam riparian vegetation, the family, genera and species of post-dam riparian vegetation survey were decreased to 73.49%, 70.41% and 57.04%, respectively. This suggests that most pre-dam riparian vegetation were unable to adapt to the opposition of natural flood rhythms, increased flooding durations, and water level fluctuation with a 30 m magnitude. Thus, plants gradually disappeared or died.

The comparative analysis of life forms of the vascular plant species indicated that the new hydrological regime imposed the most serious effects on shrub and perennial herb species. The survival and recovery of perennial species during or after flooding stress mainly depended on their root systems. The root systems of plant species surviving in the post-dam riparian zone must endure oxygen deficiency and low temperature resulting from winter flooding. In addition, these plant species must adapt to the dryer conditions in summer. Although the limited growth season resulting from increased flooding durations in the new riparian zone prevents certain annual species from completing their life cycles [2], their survival and emergence depends on seed germination [13]. Seeds may be dispersed from other places by wind or water flow along the riverbank [14].

The cluster analysis results indicated that the plant species distribution among different sampling regions had differences. Differences in plant species distribution between the Chuanjiang sections (from Chongqing to Fengjie) and Gorge sections (from Fengjie to Zigui) may be due to their different geographical environments. In the Chuanjiang sections, the river is wide, and the speed of the water flow is slow. Therefore, majority of plant seeds could be left in the WLFZ when the water level degrades. In addition, gentle slope and fertile soil were found everywhere in the WLFZ of the Chuanjiang sections. All these factors help in vegetation development. However, in the Gorge sections, the river is narrow, and the speed of water flow is fast. Therefore, few plant seeds could be left in the WLFZ when the water level degrades. In addition, steep slope and sandstone were found everywhere in the WLFZ of the Gorge sections. These factors go against vegetation development. The vegetation located in untraversed sites and experimental bases was different from the others because of rare human disturbances and advantage circumstances, or revegetation activities.

In the WLFZ of the TGA, the plant diversity is determined by the flood timing, flood duration and illumination intensity. Due to the water level of the TGR was artificially regulated within the altitude of 145 to 175 m [4], riparian vegetation in the lower section of the WLFZ encountered the most serious flood environment, whereas riparian vegetation in the upper section of the WLFZ encountered the slightest flood environment. The relative low S, H′ and D in the lower section indicate that only a few plant species could endure the serious flood environment. However, a relative high PIE in the lower section indicates that the plant community in the lower WLFZ was more unique than the communities in the upper and middle WLFZ.

Beta diversity can be defined as describing the extent of species replacement or biotic change along environmental gradients [8]. Significant differences in β diversity between lower section and other two sections, which indicated high environment heterogeneity was existed between lower section and other two sections. No significant differences in β diversity between middle section and upper section indicate that environment homogeneity was existed between middle section and upper section. High environment heterogeneity between lower section and other two sections was due to the most serious flood stressed environment in the lower section of WLFZ.

Most post-dam riparian vegetations were unable to adapt to the opposition of natural flood rhythms, increased flooding durations, and water level fluctuation with a 30 m magnitude. Thus, majority plant species will gradually disappear or die in the future. Therefore, information on selecting local plant species with tolerance to winter flooding is very useful for the recovery and re-establishment of the riparian vegetation in the WLFZ of the TGR. Plants selected for revegetation activities should be from remnant vegetation patches in the TGRA, rather than plants of the same species distributed outside the TGRA because of important local gene pools [2]. Fortunately, some plants from the TGRA showed high morphologic and physiologic plasticity to winter flooding and developed stable communities in the revegetation activities in the TGR WLFZ, including *Paspalum paspaloides*, *Cynodon dactylon*, *Vetiveria zizanioides*, *Hemarthria altissima*, *Capillipedium assimile*, *Artemisia atrovirens*, *Arundinella anomala*, *Polygonum perfoliatum*, *Distylium chinesis*, *M. laxiflora*, *Lycium chinense*, *Salix variegate*, *Salix babylonica*, *Populus nigra*, *Morus alba*, and *Taxodium ascendens* [15]. Meanwhile, reasonable distribution patterns of herb species and shrub species, as well as arbor species should be given attention during revegetation activities. These species will contribute to the spatial heterogeneity of the riparian WLFZ. Considering the impounded conditions and the seasonal hydrological regimes of the TGR, revegetation within the TGR WLFZ may be divided into three sections, including areas with lower altitudes (145 m to 160 m), middle section of 160 m to 170 m), and upper section of 170 m to 175 m. The revegetation of the lower areas should choose perennial creeping herb species with the strongest flood tolerance, including *C. dactylon*, *P. paspaloides*, *A. anomala*, and *P. perfoliatum*. The revegetation of the moderately elevated may choose perennial herb species with stronger flood tolerance and higher biomass aside from the aforementioned creeping herb species, including *H. altissima*, *V. zizanioides*, *C. assimile*, and *A. atrovirens*. These plants may be often used as raw materials of pastures, pulps, and other energy products, and revegetation using these plants may improve local economic development. The revegetation of high altitude areas should use herbs, shrubs, and arboreal species. The planted species may include shrub species and arboreal species with certain flooding tolerance, aside from the aforementioned herb species, as well as the shrub species *D. chinensis*, *M. laxiflora*, *L. chinense*, and *S. variegate*, and arboreal species *S. babylonica*, *P. nigra*, *M. alba*, *P. stenoptera*, and *T. ascendens* [16-18].

In the future, ecological succession of plant community structure maybe have a tendency that the lower section of WLFZ had the highest plant community structural stability and the middle WLFZ had the lowest. In the upper WLFZ, the infrequent flooding stress was beneficial to a wide variety of species, allowing them to survive and compete for space. Highly competitive weeds dominated these communities. In the lower WLFZ, the surviving species were those able to deal with the strongest flooding stress. The plants were highly flood stress tolerant species which could survive and form a dominant community. The flooding stress in the middle WLFZ is intermediate between that of the upper and lower WLFZs. These mid-range WLFZ species suffered from both competitive and flooding stresses. These plant species possessed a strong competitive capacity and a strong flood tolerance and formed a dominant community. The stability of this mid-zone plant community was relatively weak, and groups of plants formed co-dominant communities. Under current conditions, the plant community in the lower WLFZ was more unique than the communities in the upper and middle WLFZs. However, it is estimated the plant community compositions in the upper and middle WLFZs will gradually stabilize and become unique as the water level stabilizes around 175 meters.

Acknowledgements
This work was sponsored by National Natural Science Foundation of China (No. 31270449; 31660165), and Scientific Research Starting Foundation of Hainan University to Fan Yang (kyqd1573). We thank Mr. Zieng Zhao for his valuable contributions to the identification of plant specimens. However, any opinions, findings, conclusions, and recommendations expressed here are those of the authors, and do not reflect the official and funding agencies' views.

Reference

[1] Gleick P H 2008 Three Gorges Dam Project, Yangtze River, China. *The World's water* **2009** 139-150.
[2] New T and Xie Z 2008 Impacts of large dams on riparian vegetation: applying global experience to the case of China's Three Gorges Dam. *Biodivers. Conserv.* **17** 3149-63
[3] Zhang Q and Lou Z 2011 The environmental changes and mitigation actions in the Three Gorges Reservoir region, China. *Environ. Sci. Policy* **14** 1132–38
[4] Yang F, Wang Y and Chan Z 2015 Review of environmental conditions in the water level fluctuation zone: Perspectives on riparian vegetation engineering in the Three Gorges Reservoir, *Aquat. Ecosyst. Health* **18** 240-9.
[5] Wu J, Huang J, Han X, Gao X, He F, Jiang M, Jiang Z, Primack R B and Shen Z 2004 The Three Gorges Dam: an ecological perspective. *Front. Ecol. Environ.* **2** 241-8.
[6] Wu J, Huang J, Han X, Xie Z and Gao X 2003 Ecology: Three-Gorges Dam--experiment in habitat fragmentation? *Science* **300** 1239-40.
[7] Pitkänen S 1998 The use of diversity indices to assess the diversity of vegetation in managed boreal forests. *Forest Ecol. Manage*. **112** 121-37.
[8] Lu T, Ma K N H, Fu B, Zhang J and Lu Q 2008 Variation in species composition and diversity

of wetland communities under different disturbance intensity in the Sanjiang plain. *Acta Ecol. Sinica* **28** 1893-1900.

[9] Naiman R J and Decamps H 1997 The ecology of interfaces, riparian zones. *Ann. Rev. Ecol. System.* **28** 621-58.

[10] Nilsson C and Svedmark M 2002 Basic principles and ecological consequences of changing water regimes: riparian plant communities. *Environ. Manage.* **30** 468-80

[11] Nilsson C, Jansson R and Zinko U 1997 Long-term responses of river-margin vegetation to water-level regulation. *Science* **276** 798-800.

[12] Wang Y, Li E H and Wu J Q 2002 A preliminary study on the vascular plant flora of the water-level-fluctuating zone in the Three-Gorge Reservoir area. *J. Wuhan Bot. Res.* **20** 265-74 (in Chinese with English abstract).

[13] Lu Z J, Li L F, Jiang M X, Huang H D and Bao D C 2010 Can the soil seed bank contribute to revegetation of the drawdown zone in the Three Gorges Reservoir Region? *Plant Ecol.* **209** 153-65

[14] Chen F Q and Xie Z Q 2007 Reproductive allocation, seed dispersal and germination of *Myricaria laxiXora*, an endangered species in the Three Gorges Reservoir Area. *Plant Ecol.* **191** 67-75

[15] Yang F, Liu W, Wang J, Liao L and Wang Y 2012 Riparian vegetation's responses to the new hydrological regimes from the Three Gorges Project: Clues to revegetation in reservoir water-level-fluctuation zone. *Acta Ecologica Sinica* **32** 89-98.

[16] Cao F L and Conner W H 1999 Selection of flood-tolerant *Populus deltoides* clones for reforestation projects in China. *Forest Ecol. Manage.* **117** 211-20.

[17] Francis R A, Gurnell A M, Petts G E and Edwards P J 2005 Survival and growth responses of *Populus nigra*, *Salix elaeagnos* and *Alnus incana* cuttings to varying levels of hydric stress. *Forest Ecol. Manage.* **210** 291-301.

[18] Yang F, Wang Y, Wang J, Deng W and Liao L 2011 Different eco-physiological responses between male and female *Populus deltoides* clones to waterlogging stress. *Forest Ecol. Manage.* **262** 1963–71.

Characteristics of Summer Runoff Variations in the Yangtze River Basin and its Connections with Asian Summer Monsoons

Jian Tang[1, 2], Qingyun Li[1, 2] and Jin Chen[1, 2]*

[1] Basin Water Environmental Research Department, Changjiang River Scientific Research Institute, Wuhan 430010, China;
[2] Key Laboratory of Basin Water Resource and Eco-environmental Science in Hubei Province, Changjiang River Scientific Research Institute, Wuhan 430010, China

E-mail: Chenjincrsri@163.com

Abstract. In this research, characteristics of summer runoff variations during monsoon season in the Yangtze River basin and their connections with Asian summer monsoons were detected by using continuous wavelet transform and wavelet coherence analysis methods. The following results were obtained: (i) The periodic character of summer runoff along the mainstream of the Yangtze River basin has distinct differences upstream and downstream of Yichang station. (ii) Summer runoff of the source and upper regions of the Yangtze River basin is mainly influenced by the South Asian summer monsoons. Summer runoff of the midstream and downstream regions of the Yangtze River basin are controlled by the East Asian summer monsoons. (iii) An in-phase relationship between summer runoff and South Asian summer monsoon intensity indices is detected in the source region of the Yangtze River basin, and an anti-phase relationship between summer runoff and East Asian summer monsoon intensity indices dominates in the midstream and downstream regions of the Yangtze River basin.

1. Introduction

More than 60% of the world's population depends on Asian summer monsoon rainfall, giving this phenomenon a crucial role in providing water resources for agriculture, industrial development, and basic human needs [1, 2]. Given the reliance of lives and economies of many countries on Asian summer monsoon rainfall, understanding how Asian summer monsoons affect hydrological processes is an imperative of both scientific and societal importance [3-5].

The approximately 6380 km Yangtze River is the longest river in Asia and the third longest river in the world. Asian summer monsoons are heat conveyors and rain-bearing systems of the Yangtze River basin. The monsoons determine the runoff changes of the region [6-8]. Therefore, research on runoff variations in the Yangtze River basin and their connections with Asian summer monsoons can improve our understanding of hydro-meteorological interrelations and their implications for river dynamics and water resources management. Currently, research has focused on detecting possible connections between the El Niño–Southern Oscillation and runoff in the Yangtze River basin through statistical analysis and numerical modeling [9-11]. These studies are helpful for understanding and controlling flood and drought problems in the Yangtze River basin. However, Asian summer monsoons impacts on runoff in the Yangtze River basin remain understudied. As such, the aim of this study is to explore characteristics of average annual summer runoff variations during the Asian summer monsoon season in the Yangtze River basin, and detect the connections between average annual summer runoff and Asian summer monsoons from 1957 to 2012. The continuous wavelet transform method was employed to analyze the periodic characteristics of average annual summer runoff and Asian summer monsoons intensity indices. The relationships between summer runoff and Asian summer monsoons intensity indices were investigated with wavelet coherence analysis methods.

2. Study area and data

Summer runoff data from Zhimenda, Yichang, Hankou, and Datong hydrological stations are analyzed in this study. These four hydrological stations are used to represent the source, upper, middle, and lower reaches of the river (figure 1). Runoff data for the selected stations were collected from the hydrological yearbooks of China. Runoff data were quality controlled, and provide a continuous record from 1957 to 2012. Summer runoff data during the Asian summer monsoon season were averaged runoff over June-July-August (JJA) during the period. The Asian summer monsoons are composed of two subsystems, East Asian summer monsoons and South Asian summer monsoons [12]. Asian summer monsoons indices data were collected from http://www.lasg.ac.cn/staff/ljp.

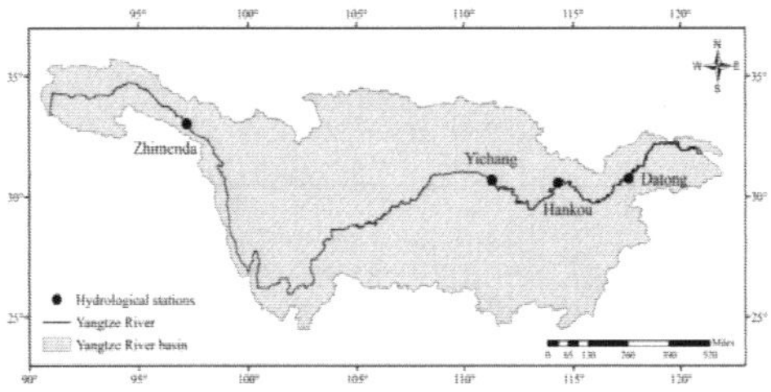

Figure 1. Sketch map of the study area.

3. Methods

In this study, the relationships between summer runoff and Asian summer monsoons intensity indices were explored by cross-wavelet and wavelet coherence analysis. The statistical confidence level of the wavelet coherence spectrum is determined by a Monte Carlo method. In order to estimate the mean and confidence interval of the phase difference, the circular mean of the phase for those regions with higher than 5% statistical significance and which are outside the cone of influence were used to quantify the phase relationship[13, 14].

4. Results and discussion

4.1. Characteristics of summer runoff variations in the Yangtze River basin

Summer runoff variation characteristics have distinct differences between the upper and lower reaches of the Yangtze River. For the Zhimenda and Yichang stations located in the upper reaches, high wavelet power in the 1- to 3-year band is detected in the wavelet power spectra for summer runoff around 1991~1995. In the Hankou and Datong stations located in the lower reaches, high wavelet power in the 7- to 10-year band is detected in the wavelet power spectra for summer runoff around 1972~1990(figure 2).

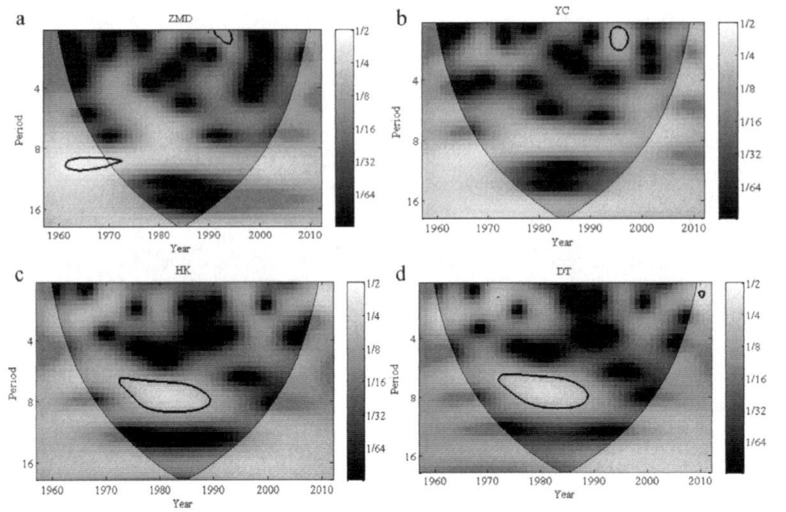

Figure 2. Power spectra of continuous wavelet transform for normalized summer runoff time series in the Yangtze River basin at (a) Zhimenda, source reach; (b) Yichang, upper reach; (c) Hankou, middle reach and (d) Datong stations, lower reach.

4.2. Connections between summer runoff and Asian summer monsoons

To elaborate the relationship between Asian summer monsoons and summer runoff in the Yangtze River basin, wavelet coherence analysis is applied to detect the links between the two time series. Wavelet coherence results show a coherence phase in the 3- to 5-year band around 1986~1995 and 2000~2005. These regions show an in-phase relationship between summer runoff and East Asian summer monsoon intensity indices (figure 3a). From 1990 to 1998, an anti-phase relationship between summer runoff and South Asian summer monsoon intensity indices is detected in the 2- to 6-year band (figure 3b).

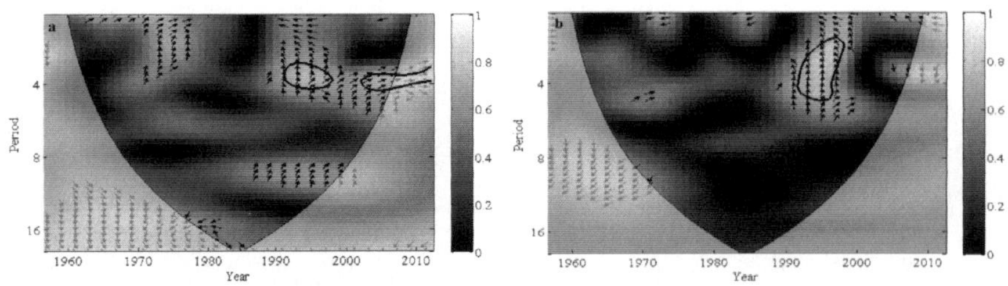

Figure 3. Wavelet coherence spectra for summer runoff at Zhimenda station and Asian summer monsoon intensity indices.

A coherence phase of 1- to 5-year band around 1980~1990 and 1992~1998 is detected in the wavelet coherence power spectra. These regions show anti-phase relationships between summer runoff and East Asian summer monsoons. An in-phase relationship between summer runoff and East Asian summer monsoons is detected in the 4- to 7-year band around 1965~1975 (figure 4a). The wavelet coherence for summer runoff and South Asian summer monsoons shows there are relationships in the 1- to 6-year band around 1963~1971 (in-phase relationship), 1980~1990 (in-phase relationship), and 1991~2002 (anti-phase relationship) (figure 4b).

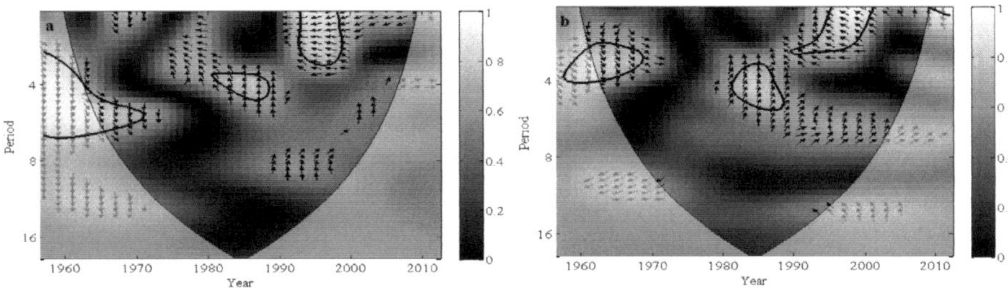

Figure 4. Wavelet coherence spectra for summer runoff at Yichang station and Asian summer monsoon intensity indices.

Wavelet coherence results show a coherence phase in the 5- to 12-year band around 1963~1973 and 1990~2005. Phase relationships between summer runoff and East Asian summer monsoon intensity indices are unstable in these regions. There is also an anti-phase relationship between summer runoff and East Asian summer monsoon intensity indices in the 1- to 4-year band around 1969~1995 (figure 5a). A coherence phase in the 2- to 5-year band around 1961~1969 is detected in the wavelet coherence power spectra for summer runoff and South Asian summer monsoon intensity indices. There is an in-phase phase relationship between summer runoff and South Asian summer monsoon in this region (figure 5b).

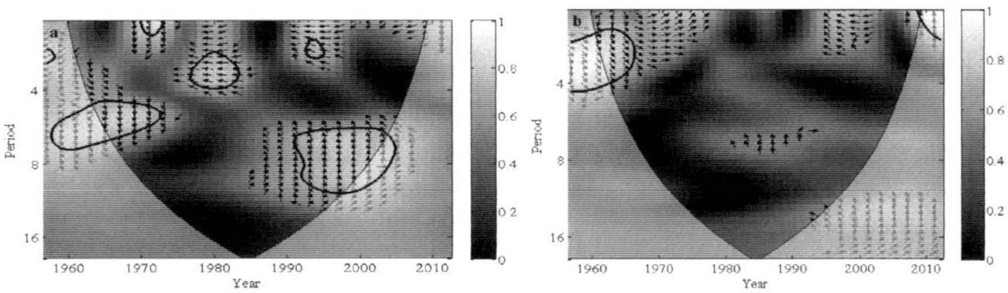

Figure 5. Wavelet coherence spectra for summer runoff at Hankou station and Asian summer monsoon intensity indices.

The phase relationships between summer runoff and East Asian summer monsoon intensity indices in the region of 1963~1973 are unstable. Phase relationships between summer runoff and East Asian summer monsoon intensity indices in the regions of 1978~1985 and 1991~1993 are anti-phase. There also is a coherence phase in the 5- to 12-year band around 1989~2005. The phase relationships between summer runoff and East Asian summer monsoon intensity indices are unstable in this region (figure 6a). The wavelet coherence for summer runoff and South Asian summer monsoon intensity indices shows there is a coherence phase in the 1- to 5-year band around 1962~1964 and 1970~1978. There is an in-phase relationship between summer runoff and South Asian summer monsoon intensity indices in this region (figure 6b).

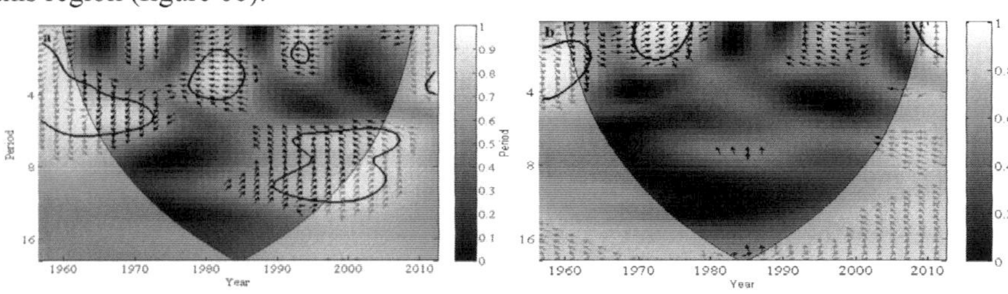

Figure 6. Wavelet coherence spectra for summer runoff at Datong station and Asian summer monsoon intensity indices.

The wavelet coherence spectra results indicate that the meridional water vapor fluxes carried by the East Asian summer monsoons are dominant in the midstream and downstream regions of the Yangtze

River basin. In contrast, the zonal water vapor fluxes are relatively large in the source and upper regions of the Yangtze River, located in arid and semiarid regions of North-western China [15]. An anti-phase relationship between summer runoff and East Asian summer monsoon intensity indices dominates in the midstream and downstream regions of the Yangtze River basin (figure 5 and 6). In these regions, weak East Asian summer monsoons will result in floods, and relatively strong East Asian summer monsoons are usually associated with droughts [16]. In the source region of the Yangtze River basin, an in-phase relationship between summer runoff and south Asian summer monsoon intensity indices is detected. Phase relationships between summer runoff at Yichang station and Asian summer monsoons are ambiguous, and have in-phase and anti-phase relationships simultaneously.

5. Summary

With the help of continuous wavelet transform and wavelet coherence analysis methods, this research explored the characteristics of summer runoff variations along the mainstream of the Yangtze River basin and its connections with Asian summer monsoons from 1957 to 2012. The following conclusions were drawn:

(1) There are obvious differences in periodic characteristics of summer runoff along the mainstream of the Yangtze River basin.

(2) Summer runoff of the source and upper regions of the Yangtze River basin is mainly influenced by zonal water vapor fluxes carried by South Asian summer monsoons. Summer runoff of the midstream and downstream regions of the Yangtze River basin is controlled by meridional water vapor fluxes carried by East Asian summer monsoons.

(3) Anti-phase relationships between summer runoff and East Asian summer monsoons dominate in the midstream and downstream regions of the Yangtze River basin, and in-phase relationships between summer runoff and South Asian summer monsoons are detected in the source region of the Yangtze River basin. Phase relationships between summer runoff at Yichang station and Asian summer monsoons are ambiguous.

Acknowledgments

This study was supported and funded by National Natural Science Foundation of China (Grants No. 51609008); China Postdoctoral Science Foundation (Grants No. 2015M572124), and China CDM Fund Project (Grants No. 2012044).

References

[1] Turner A G and Annamalai H 2012 Nat. *Climate Change* **2** 587-595
[2] Wu G X, Liu Y M, He B, Bao Q, Duan A M and Jin F F 2012 *Sci. Rep.* **2** 404
[3] Asharaf S, Dobler A and Ahrens B 2012 *J. Hydrometeor.* **13** 1461-1474
[4] Lu H Y et al. 2013 *Geology* **41** 1023-1026
[5] Sun Y B et al. 2015 *Quaternary Sci. Rev.* **115** 132-142
[6] Blender R, Zhu X H, Zhang D and Fraedrich K 2011 *Quatern. Int.* **244** 194-201
[7] Zeng G, Wang W C, Shen C M and Hao Z X 2014 J. *Atmos. Sci.* **50** 355-364
[8] Kubota Y, Tada R and Kimoto K 2015 *Climate Past* **11** 265-281
[9] Zhang Q, Xu C Y, Jiang T and Wu Y 2007 *J. Hydrol.* **333** 265-274
[10] Hu K M, Huang G and Wu R G 2013 *J. Climate* **26** 2205-2221
[11] Zhang Z Z, Chao B F, Chen J L and Wilson C R 2015 *Global Planet. Change* **126** 35-45
[12] Wang Y Q and Zhou L 2005 *Geophys. Res. Lett.* **32** doi:10.1029/2005GL02257
[13] Torrence C and Compo G P 1998 Bull. *Amer. Meteor. Soc.* **79** 61-78
[14] Sang Y F 2013 *Atmos. Res.* **122** 8-15
[15] Huang R H, Chen J L, Wang L and Lin Z D 2012 *Adv. Atmos. Sci.* **29** 910-942
[16] Qian Y F, Zheng Y Q, Zhang Y and Miao M Q 2003 *Int. J. Climatol.* **23** 593-613

Sustainable Improvement of Urban River Network Water Quality and Flood Control Capacity by a Hydrodynamic Control Approach-Case Study of Changshu City

Chen XIE[1,2]**, Fan YANG**[1,2]**, Guoqing Liu**[1,2]**, Yang LIU**[1,2]**, Long WANG**[1,2] **and Ziwu Fan**[1,2]

[1]Hydraulic Engineering Department, Nanjing Hydraulic Research Institute, Nanjing 210029, China;
[2]State Key Laboratory of Hydrology-Water Resources and Hydraulic Engineering, Nanjing Hydraulic Research Institute, Nanjing 210029, China

E-mail: 19434455@qq.com, yangfan-1992@foxmail.com

Abstract. Water environment of urban rivers suffers degradation with the impacts of urban expansion, especially in Yangtze River Delta. The water area in cites decreased sharply, and some rivers were cut off because of estate development, which brings the problems of urban flooding, flow stagnation and water deterioration. The approach aims to enhance flood control capability and improve the urban river water quality by planning gate-pump stations surrounding the cities and optimizing the locations and functions of the pumps, sluice gates, weirs in the urban river network. These gate-pump stations together with the sluice gates and weirs guarantee the ability to control the water level in the rivers and creating hydraulic gradient artificially according to mathematical model. Therefore the flow velocity increases, which increases the rate of water exchange, the DO concentration and water body self-purification ability. By site survey and prototype measurement, the river problems are evaluated and basic data are collected. The hydrodynamic model of the river network is established and calibrated to simulate the scenarios. The schemes of water quality improvement, including optimizing layout of the water distribution projects, improvement of the flow discharge in the river network and planning the drainage capacity are decided by comprehensive Analysis. Finally the paper introduces the case study of the approach in Changshu City, where the approach is successfully implemented.

1. Introduction

The Yangtze River Delta has favorable natural conditions, good geographical conditions, solid economic base, advanced science, technology and culture education, with convenient land and water transportation. It is the most economically advanced region in China. There are complex river network, abundant precipitation, and rich water resources. Water environment of urban rivers suffers degradation with the impacts of urban expansion[1]. Some urban river channels were occupied in the process of urbanization. A large number of sluice gates make the urban river network be isolated. The sludge and contaminants in the urban river continuously deposited. Few of water functional areas could meet the water quality standard. Many urban rivers were polluted and the aquatic environment deteriorated[2-3]. Therefore, the water environment of the river network regions to be improved urgently.

The purpose of the hydrodynamic control approach for urban flood control and water environment improvement is to apply the solutions for the occupied city river, the deterioration of the water environment, the slowly velocity of river network, poor water quality and other difficult problems in Yangtze River Delta. Urban river water environment is an important element for peoples life. The river system in Changshu City is typical urban river network in Yangtze river delta[4]. It faced all the above problems and additional flood control problem. The paper inroduces the hydrodynamic control approach which combining urban flood control with water environment development. It can realize almost all of the rivers flow smoothly in Changshu city, the reliable protection of the flood containment and the improvement of the river water ecological system.

2. Methodology

The approach can effectively improve the urban water environment and enhance urban flood control capacity. The basic ideas and the key techniques are listed below.

2.1 Basic ideas of improvement the urban river network

Step 1: Identify the clean water sources.

The water sources should be sufficient and the water quality should be much better than that in the urban rivers, i.e. the water in Yangtze river. Also, the selection of the water sources is according to its locations in the drainage basin and regional river network. The water sources are diverted into city by pump station or by water level differences.

Step 2: Water intake and outlet routes.

The selection is according to the landform, river system pattern and the existing urban clean water channel. There are usually more than one water intakes and outlet. The major rivers are usually chosen as the intake or outlet channels. The water level in the intake channel is higher than the outlet. The water level differences are created by the existing gates and weirs artificially. The pump-sluice stations are planned in the intake and outlet, which are helpful for flood control and create the hydraulic gradient.

Step 3: Estimation of the water-recycling rate in demand in the urban river network.

The maximum water-recycling rate of the river network can be estimated according to the urban river storage capacity and the capacity of the pump for water intake. The total urban river storage capacity can be calculated by the hydrodynamic model.

Step 4: Making the river flow smoothly by creating water level difference artificially.

The water levels in the river are planned and arranged according to the hydrodynamic model. Sluice gates and weirs are controlled to create water level differences. Avoiding to use the pumps in the urban area. Some weirs may be built to control if there is no existed gate. Water flow smoothly by the hydraulic gradient.

Step 5: Improvement of water quality sustainably.

The flow velocity increases in the urban rivers, which increases the rate of water exchange, the DO concentration and water body self-purification ability. The water quality improves day by day.

Step 6: Improvement of the flood control capacity.

For some cities, there are no gate-pump station around the city. The city may be suffered flood by heavy rainfall. The stations are planned in the outlet of the rivers around of the city to improvement of the flood control capacity. During the non-flood season, these stations can be used to intake clean water or making the water level differences.

2.2 Key techniques

The above ideas are realized by the following key techniques [5].

(1) Site survey. Assessing the conditions and problems of the gate-pump station and rivers by site survey. There are some common problems in the urban rivers. The water almost does not flow in the rivers. Water quality looks not good. Water blackening and stink in some rivers. Some rivers are not

connected. The date of gate-pump stations are different from the books. The scheduling operation of these stations are not intelligent, etc.

(2) Prototype observation. The water level, flow rate and water quality of the urban river are observed synchronously. These data can be used for calibration and verification of the hydrodynamic model. And it also can provide the information for analysis.

(3) Hydrodynamic model. The one-dimensional hydrodynamic model is constructed by InfoWorks RS or InfoWorks ICM, and the model is calibrated with the measured data. The error can be reduced to less than 2cm, and the model can be used to simulate the scenarios closing to the reality.

(4) Scheme analysis. Scenarios are simulated, including the status of the gate-pump station, water level, discharges, velocity, the raising of water level when flooding, etc. A proper scheme for the urban river network are determined. All the rivers can flow or enhance the flow velocity in a certain district.

(5) Site test. According to the scheme, a site test can be conducted by controlling the existing projects and some temporary facilities.

3. Case study of Changshu city

3.1 The current situation of water system

Changshu is under the jurisdiction of Suzhou, in Jiangsu province. It located in the north of Suzhou City, east of Wuxi city and south of Yangtze river. Total area of Changshu is 1264 km^2. The main urban area is about 60 km^2, where is the case study area. There are typical river network.Changshu belongs to the Taihu Lake basin. The key rivers are Haiyangjing River, Baimaotang River, Changxu River, Yuanhetang River, Genjingtang River and so on.

There are over 150 rivers, 99 gate or/and pump stations in the urban area and the total length of the river is 159.3km, as shown in Figure 1. There is Mount Yu in the east, Shanghu Lake and Kuncheng Lake in the south. The major problems of the rivers are listed as bellow,

a) Rivers in Changshu are radial water system. The ancient city are in the centre about 2 km^2. There are 10 divergent key rivers and 3 circular key rivers, as shown in Figure 2. Therefore it is difficult to make all the water in the river flow smoothly.

b) The rivers are lack of external control engineering facilities.

c) It is hard to pump the water into all the urban rivers. Because the large proportion water flows directly from the main river.

d) The river water is turbidity and the sense is poor, because the dredging is not complete and the water is polluted.

e) A large number of the urban rivers are lost or disconnected during the city expansion process.

Figure 1 Urabn river network of Changshu (red points are gates and pumps)

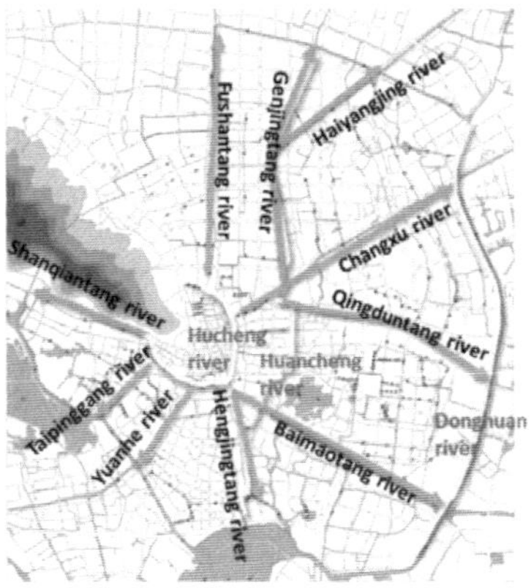

Figure 2 Key rivers of Changshu city

3.2 Sustainable scheme to improve the urban river network of Changshu

The hydrodynamic model of Changshu consists of 1135 river sections, 790 river reaches adding up to 129km and 281 cross points, which is constructed by the InfoWorks RS software. In the software the gates and pumps are processed with linear object and the corresponding buildings are constructed between the sections. Then the building size information and the corresponding rules should be given in the software. The sub watersheds are determined by the scope of the river boundary and use the side stream object to associate the watershed to the river sections. The model is constructed to create a total of 93 catchments, as shown in Figure 3. The planning scheme in Changshou City consists 3 gates, 7 gate-pump stations and 5 rotated weirs, with the total capacity of $170m^3/s$, as shown in Figure 4. The gate-pump faculties are used to control the flood in the flood seasons and to improve the water quality in the non-flood seasons. In the non-flood seasons the gate-pumps are used to pump the fresh water from Yangtze River into the Changshu. The main water diversion channel is Haiyang River, with good and stable water quality, and the maximum pump diversion flow is $30m^3/s$. Minor diversion channel is Shanqian River. When the Wangyu River water quality is good, Shanqian River is used as a fresh channel which can provide the flow of $12\text{-}13m^3/s$. 5 rotated weirs are planned to enhance the control ability in the city. Together with the planned flood control facility, all the rivers in the city can be controlled artificially. Schemes for the whole and part of the urban area are simulated.

Figure 3 Hydrodynamic mathematical model of Changshu city

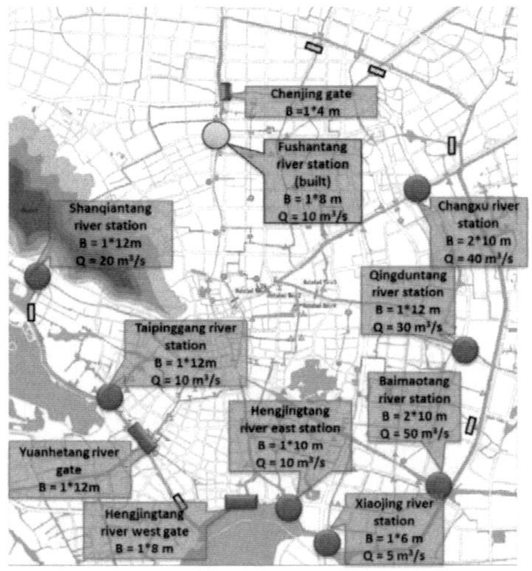

Figure 4 Planning flood control projects (projects can also be used to improve the water environment)

3.3 The implementation effects

Through the establishment of flood control area and control of the regional gates, the following effects can be achieved: a) Flood control standard can be add up to 100 years, the drainage standard can be add up to 20 years. b) Almost all of the rivers can flow. Especially, the flow can increase to $1.5\sim2.0m^3/s$ in the ancient City region, and the replacement of the whole water can be 2-3 times a day in ancient City region, 3-4 times a day in other areas. c)It can eliminate the black and smelly water in the river completely and improve the river water quality to meet the basic landscape water standard comprehensively.

4.Conclusion

This study shows a hydrodynamic approach to improve the urban river network based on hydrodynamic mathematical model. The hydrodynamic model is one the most significant tools for the improvement of urban water environment and the governance of urban waterlogging. In addition, urban river network hydrodynamic model can accurately simulate the hydraulic characteristics, such as, water level, flow velocity, and flow pattern of urban river. Meanwhile, it was utilized to analyze the scientificity and rationality of the project, which provided great help on the formulation and implementation of plans

This study reasonably improved the flow velocity and adjusted the water level by use of the existing gate-pumps and some planning projects, creating hydraulic gradients. It realized the comprehensive control of urban water environment.

The self-adjustment and self-maintenance of urban water system can be realized through the hydrodynamic control approach which can both improve urban water environment and manage urban waterlogging at the same time. Therefore, the hydrodynamic control approach, which combined with city flood protection and water environment promotion, is sustainable, economical, and worth popularizing and applying.

Acknowledgement

The study is financially supported by the National Key Research and Development Program of China（No. 2016 YFC0401500）, 7th sub-topic (No. 2016YFC0401507), 3rd sub-topic (No. 2016YFC0401503), the National Natural Science Foundation of China (Grant No. 41402217), and the Fundamental Research Funds of the National Non-Profit Research Institutes of NHRI (No. Y114010).

References

[1] ZHU Dakui, WANG Ying, WANG Dong, et al. 2004 Study on water resources and water environment On the delta plain of the changjiang river. *Quaternary Sciences*, 24 (5): 486-494. (In Chinese)

[2] LI Shuangcheng, ZHAO Zhiqiang, WANG Yanglin. 2009 Urbanization Process and Effects of Natural Resource and Environment in China: Research Trends and Future Directions. *Progress in Geography*, 28 (1): 63-70. (In Chinese)

[3] BAI Yiqin. 2010 *Study on the Characteristics of river network changes and its Protection under Rapid Urbanization in Pudong New Area*. East China Normal University. (In Chinese)

[4] FAN Ziwu, Zhou E, LI Yun, XIE Chen, et al. 2016 *Study on the planning of flowing water in Changshu city*. Nanjing: Nanjing Hydraulic Research Institute: 7-17. (In Chinese)

[5] FAN Ziwu, ZHAO Ruilong, LI Yun, et al. 2016 *Study on improvement of urban river network water quality in Suzhou ancient city by a hydrodynamic control approach*. Nanjing: Nanjing Hydraulic Research Institute: 7-17. (In Chinese)

Capabilities of Seven Species of Aquatic Macrophytes for Phytoremediation of Pentachlorophenol Contaminated Sediment

Liangyuan Zhao, Weijie Guo, Qingyun Li, Huan Li, Weihua Zhao and Xiaohuan Cao

Basin Water Environmental Research Department, Yangtze River Scientific Research Institute, Wuhan 430010, China

zhaoliangyuannew@163.com

Abstract. Sediments are regarded as the ultimate sink of pentachlorophenol(PCP) in aquatic environment, and capabilities of seven species of aquatic macrophytes for remediating PCP contaminated sediment were investigated. Seven species of aquatic macrophytes could significantly accelerate the degradation of PCP in sediments. Among all, canna indica L., Acorus calamus L. and Iris tectorum Maxim. can be used as efficient alternative plants for remediation of PCP contaminated sediment,which attained 98%, 92% and 88% of PCP removal in sediments, respectively. PCP was detected only in root tissues and the uptake was closely related to the root lipid contents of seven plants. The presence of seven aquatic macrophytes significantly increased microbial populations and the activities of dehydrogenase compared with control sediments, indicating that rhizosphere microorganism played important role in the remediation process. In conclusion, seven species of aquatic macrophytes may act as promising tools for the PCP phytoremediation in aquatic environment, especially *Canna indica L., Acorus calamus L.* and *Iris tectorum Maxim.*

1. Introduction

Pentachlorophenol (PCP) is a polychlorinated aromatic compound, classified as persistent organic pollutants (POPs) by the Stockholm Convention [1-2]. In china, PCP has been widely used for killing oncomelania, the only intermediate host of schistosome, in the middle and lower reaches of Yangtze River where schistosomiasis is epidemic in recent decades. PCP has been classified as a group 2B carcinogen by the International Agency for Research on Cancer (IARC) [1]. Owing to its toxicity, widespread distribution and its impurities (e.g. dioxin, furans and hexachlorobenzene), PCP has been listed as a priority pollutant or restricted by many countries including China [1, 3].

Considering the potential health risks of PCP, china has restricted the production and use of Na-PCP in 1997[3]. However, with the re-emergence of schistosomiasis in the traditionally epidemic areas, the production and use of PCP for snail elimination and schistosomiasis control were warranted once again [3]. In recent years, PCP and other molluscacides were mixed used by some schistosomiasis prevention and control departments to improve the molluscicidal effect. Wide use of PCP resulting in wide occurrence of PCP at relatively high concentrations in aquatic environment [3]. For example, PCP residues in sediments from the Dongting lake, Yangtze River catchment of wuhan, Bottomland of Poyang lake, Pond of Tangxun Lake and the Miaohe in china were in the concentrations of 483000, 39.14, 80.78, 43 and 230 ng/g, respectively[3]. Thus, removal of PCP from sediment is necessary.

The emerging technology of phytoremediation, originally applied to decontamination of metals-contaminated soils, has been extended to organic contaminants thanks to the ecological and economic sustainability [4-5]. It is important to investigate the phytoremediation of PCP contaminated sediment using aquatic macrophytes, because aquatic macrophytes was more suitable for growing in the aquatic environment. However, phytoremediation studies for PCP are very limited and more aquatic plants should be screened for supplementing the phytoremediation theory.

The aim of the present investigation was to evaluate the potential and efficacy of aquatic macrophytes to remediate the sediment contaminated with PCP and seven species of common aquatic macrophytes were selected. The research results are believed to promote the development of ecological remediation of PCP contaminated environment in the polluted areas.

2. Materials and methods

2.1. Chemicals

PCP (PCP, C_6Cl_5OH) used in the experiment was of 98% chemical purity(AR) and a standard solution of PCP was of HPLC grade(dissolved in methanol).

2.2. Sediments

Samples of sediments without detectable PCP were collected from the 0-20 cm surface sediments at a local lake in Wuhan, Hubei province, China. Physical and chemical properties of sediments were displayed in the Table 1.

Table 1. Physical and chemical properties of sediments used in this study.

Parameter	Unit	Value(Mean ± SD)
pH (H_2O)	-	7.86±0.08
Total nitrogen (TN)	mg/g	0.41±0.03
Total phosphorus (TP)	mg/g	0.13±0.02
Organic matter (OM)	%	3.38±0.41

2.3. Aquatic macrophytes

Seven speciess of aquatic macrophytes, including *Canna indica L., Acorus calamus L., Alisma plantago-aquatica L., Polygonum hydropiper L., Thalia dealbata Fraser, Iris tectorum Maxim., Cyperus alternifolius L.*, used in the experiment were collected from Wuhan xiushui ecological engineering co., LTD in Wuhan, Hubei province, China.

2.4. Phytoremediation treatments setup

Sediment samples collected were air-dried, passed through a 2-mm sieve. PCP dissolved in methanol was uniformly sprayed onto a portion of uncontaminated sediment to produce a PCP level of about 40 mg/kg. A series of contaminated sediments(approximately 2 mg/kg) were prepared by mixing the spiked sediment with appropriate amounts of the uncontaminated sediments.

For the phytoremediation experiment, 10 kg of PCP contaminated dry sediments were loosely packed into planting pots (0.3m×0.4m×0.3m). 6-10 individuals of seven macrophytes with initial plant height of about 15 cm were transplanted to experimental pots at initial stage with the aim of obtaining an approximately equivalent amount of plant biomass per pot. The phytoremediation experiment setup was displayed in Figure 1.Tap water was then added and kept at 4 cm above the sediments surface. The experiment ran for 120 days.

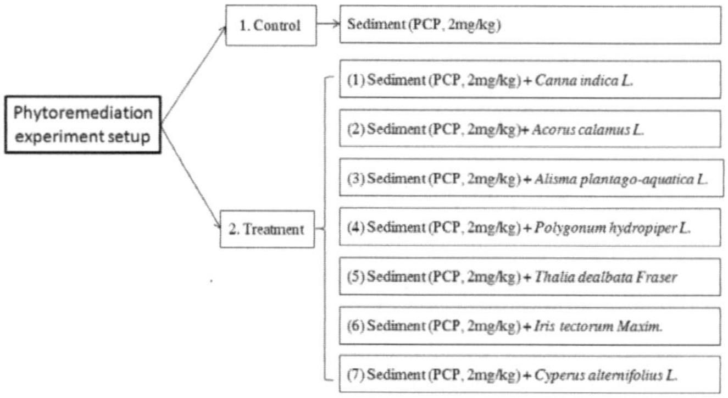

Figure 1. Phytoremediation experiment setup

2.5. Sample preparation

For the plants samples, root tissue and shoot tissue of seven species of aquatic macrophytes were randomly harvested and washed with tap water and distilled water to remove the adherent sediments particles. After that, clean samples were dried at 60°C for 24 h and preserved at -20°C in glass for further PCP analysis.

Rhizosphere sediments were sampled using sampling spade by vigorous shaking and rolling the roots to remove any remaining sediments. Then the sediments collected were oven dried, grounded, homogenized and passed through a 80 mm mesh standard sieve. Sediment samples were oven-dried, ground, and homogenized by sieving through a stainless steel 80-mesh (0.18 mm) sieve and stored in glass containers at -20 °C until extraction.

All the samples above were collected monthly.

2.6 .Chemical analysis

2.6.1. PCP extraction and cleanup. Three replicate samples of 5g (dw) each were extracted with 30 mL of methanol for 30 min by ultrasonic extraction. This extraction was performed triple. The combined 90 mL of extract was deeply filtrated by glass microfiber filters (GF/C, whatman) and then concentrated to about 3 mL by rotary evaporation at 40 °C. Then the extract was concentrated under a gentle stream of nitrogen and the final extract volume was 1.0 mL. After filtration through 0.22-μm filter units (hydrophobic), the treated sediment extracts were determinated for PCP using HPLC.

The dried samples of root tissue, shoot tissue collected in every month were separately combined to determine the PCP concentrations in the plant tissue. PCP was extracted from each plant part (5 g) by ultrasonic extraction for 30 min with 30 mL of *n*-hexane / dichloromethane (1:1, v: v) and the same extraction was performed triple. The combined extract was deeply filtrated by glass microfiber filters (GF/C, whatman) and concentrated to about 3 mL and treated with 3 mL of concentrated H_2SO_4. All the concentrated extracts were cleaned and fractionated on a 1 cm i.d. silica/ anhydrous sodium sulfate alumina column packed, from the bottom to top, with silica gel (15 cm), anhydrous sodium sulfate (15cm).The mixture was eluted with 90 ml of n-hexane: dichloromethane (1:1, v:v). The extract was concentrated and solvent-exchanged to methanol under a gentle stream of nitrogen and the final extract volume of macrophyte was 1.0 mL. After filtration through 0.22-μm filter units (hydrophobic), the treated plant extracts were determinated for PCP using HPLC.

2.6.2. PCP analysis. Quantitative analysis of PCP was performed by an HPLC (Waters Alliance e2695-2489, USA) equipped with UV spectrophotometric detector. A C18 column (4.6 mm × 150 mm I.D.; 5-μm particle size) from Waters was used. The Chromatographic condition for PCP determination was: the flow rate was 1.0 ,mL/min and the mobile phase was water : methanol (20:80), The detection wavelength was 305nm, and the column temperature was 35°C.

2.6.3. Lipid content in roots. The root tissue samples of seven species of aquatic macrophytes were dried at 60°C for 24 h, and were cut into pieces. 5g of dry samples were sonicated in a 30mL mixture of n-hexane and acetone (1:1,v:v) for 20 min at 30°C using an ultrasonic bath, and this process was replicated three times. The extract was cominbined and poured into a evaporating dish to obtain the lipid using natural evaporation method. The lipid content was weighed by analytical balance (millesimal) (in units of %).

2.7. Biological analysis.

The bacterial abundance was measured using the colony-forming units (CFUs) method. A 1-g aliquot of sediment sample was collected, serially diluted, and spread on a Luria–Bertani medium. The medium was incubated for 24h at 37 °C, and then the colonies formed were directly counted (in units of cells/g dw).

Dehydrogenase activity of sediment samples was measured by triphenyl tetrazolium chloride colorimetry[6] and was calculated as μg TPF g^{-1} sediment $24h^{-1}$ dw.

2.8. Statistical analysis.

One-way analysis of variance (ANOVA) was used to examine the significance among PCP dissipation rates, PCP accumulation in plants, microbial populations and dehydrogenase activities. Differences between two parameters were compared by independent-samples t-test (two-tailed). Pearson correlations (r_p) (2-tailed) were performed to describe the relationships among multiple measured variables. All data was presented as mean± SD (n=3) and statistical analyses were performed by using GraphPad Prism 5.0.

3. Results and discussions

3.1. Dissipation rates of PCP

Figure 2 demonstrated the dissipation rates of PCP in the control and treatment sediments planted with seven species of aquatic macrophytes. After 120 days, approximately 33% of PCP was disappeared in the control sediment, whereas seven species of aquatic macrophytes significantly facilitated dissipation rates(60%-98%) of PCP in the treatment sediments(p<0.01). Among all the aquatic macrophytes , *Canna indica L.*, *Acorus calamus L.* and *Iris tectorum Maxim.* exhibited the fastest PCP Dissipation rates, which attained the 98%, 92% and 88% of PCP removal in sediments, respectively. Thus, Those three species of aquatic macrophytes can be used as efficient alternative plants for remediation of PCP contaminated sediment.

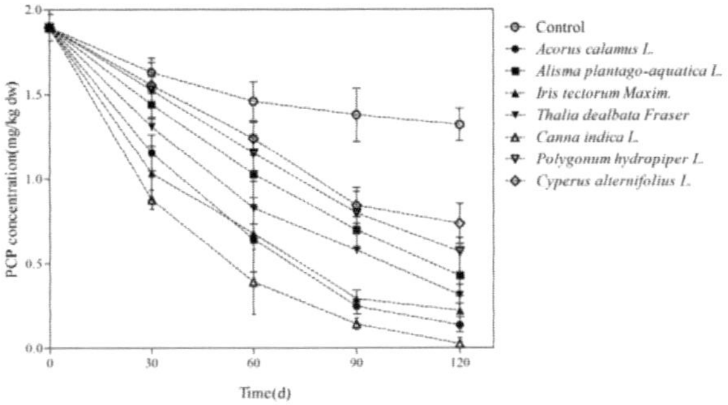

Figure 2. Changes of PCP concentrations in the control and treatment sediment

3.2. PCP uptake by macrophytes

The contribution of plant uptake and accumulation to the dissipation enhancement of organic pollutants was negligible contrary to the heavy metal remediated by plants [7-8]. PCP accumulation in seven aquatic macrophyte samples was further determined after 120 days (Figure 3). PCP was detected only in root tissues (approximately 100-500μg/kg dw), which indicated that PCP was hard to translocated up to the shoot tissue, which was accordant with the previous research [9]. Previous research demonstrated that organics that is most likely to be taken up by plants are moderately hydrophobic compounds with Log *Kow* ranging from 0.5 to 3[9]. Since the Log *Kow* of PCP was 5.01, which might make it difficult to transport from root tissue to upper part tissue.

The lipid content of root for seven species of aquatic macrophytes was detemined(Figure 4) and a significant positive correlation was obtained between the PCP concentrations in root tissues and root lipid content of seven species of aquatic macrophytes ($r=0.542$, $p<0.0001$), indicating that PCP uptake was mainly associated with lipid content of each plant. Thus, plants with root of higher lipid content is an option for screening remediation plants in the phytoremediation process of PCP contaminated sediment.

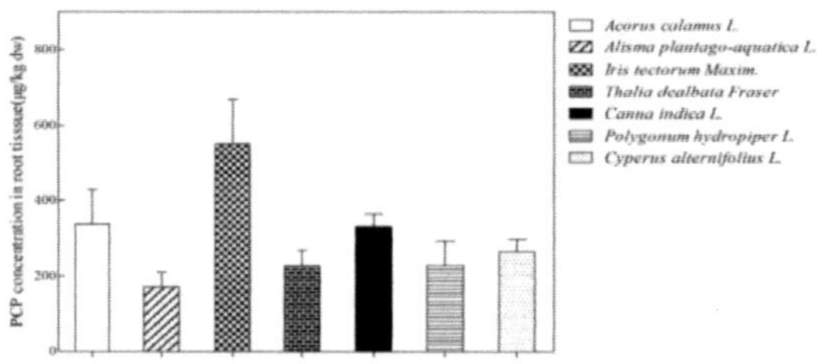

Figure 3. PCP concentrations detected in the root tissue of seven species of aquatic macrophytes.

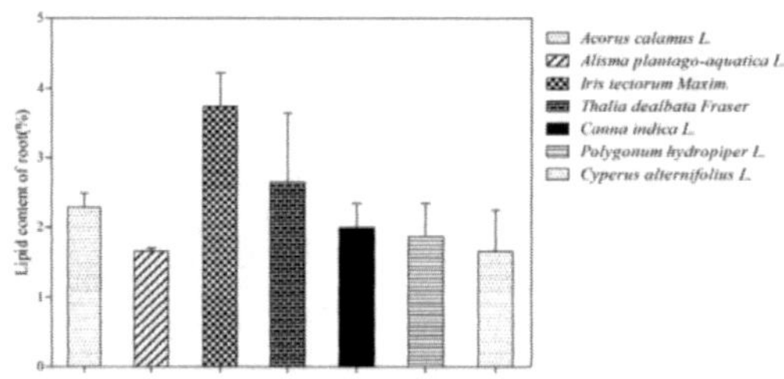

Figure 4. Root lipid content of seven species of aquatic macrophytes.

3.3. The change of microbial populations

Rhizosphere is biologically active soil region where macrophyte roots interact with sediment and microbes and roots provide suitable habitats for the growth of the microorganisms [10]. Furthermore, plant exudation and root sloughing can also affect the activities of microbes, and changes in exudates released resulting in associated effects on biodegradation and microbial populations [11-12].Thus, it is important to explore the microbiological changes in the rhizosphere of seven species of aquatic macrophytes.

Significant dissipation of PCP in the rhizosphere of seven species of aquatic macrophytes, as compared to control sediment, might be due to higher activity and density of microorganisms in the rhizosphere along with root exudates enhancing PCP availability. The microbial populations in the three sediment controls and treatment sediments are showed in Figure 5. The data revealed that microbial numbers in the planting sediment were higher than those in unplanted sediment ($p<0.01$). The microbial numbers of the rhizosphere sediment of seven species of aquatic macrophytes were almost 2-4 times more than the control sediment.

The presence of seven aquatic macrophytes also significantly increased the activities of dehydrogenase($p<0.001$) compared with control sediments(Figure 5).The above results clearly demonstrated that the microbial populations were significantly enhanced by growth of aquatic macrophytes. Furthermore, a significant relationship was obtained between the amount of microorganism (r=0.879, p<0.001) and dehydrogenase activity(r=0.898, p<0.001) in the sediments.Those indicated the microorganism in rhizosphere played important role in the remediation process.

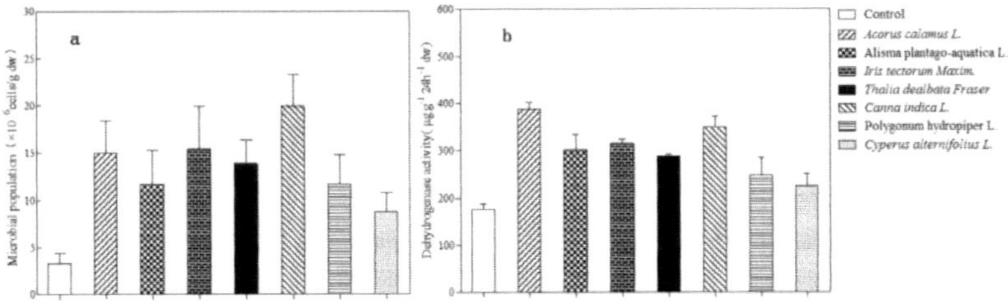

Figure 5. Microbial populations (a) and dehydrogenase activities (b) of the control and treatment sediment

4. Conclusion

The present study demonstrated the use of aquatic macrophytes for cost-effective removal of PCP from polluted sediments. The presence of seven species of aquatic macrophytes significantly increased the dissipation of PCP(60%-98%) in sediment compared with no planting. PCP was detected only in root tissues and the uptake was closely related to the root lipid contents of seven plants. Rhizosphere microorganisms played an important role in the process of phytoremediation. The results indicated that phytoremediation by aquatic macrophytes, especially *Canna indica L., Acorus calamus L.* and *Iris tectorum Maxim.* is a promising way for the PCP remediation in aquatic environment.

Acknowledgments

The research was financially supported by National Natural Science Foundation of China (Grant No.51309020) and Non-profit Industry Financial Program of MWR (Grant No.201501042, 201501019, 201401020) and Special fund for basic scientific research business of central public research institutes(Grant No. CKSF2015016).

References:

[1] Zheng W, Wang X, Yu H, Tao X, Zhou Y, and Qu W 2011 *Environ. Sci. Technol.* **45** 4668-4675
[2] Yang S, Shibataa A, Yoshida N and Katayama A 2009 *Biotechnol. Bioeng.* **102** 81-90
[3] ZhengW, Yu H, Xia W and Qu W 2012 Environ. Int. **42** 105-116.
[4] Teng Y, Shen Y Y, Luo Y M, Sun X H, Sun M M, Fu D Q and Li Z G, Christie P 2011 *J. Hazard. Mater.* **186** 1271-1276
[5] Soleimani M, Afyuni M, Hajabbasi MA, Nourbakhsh F, Sabzalian M R and Christensen J H 2010 *Chemosphere.* **81** 1084-1090
[6] Cheema S A, Khan M I, Shen C F, Tang X J, Farooq M, Chen L, Zhang CK and Chen YX 2010 *J. Hazard. Mater.* **177** 384-389
[7] Burken J G and Schnoor J L 1998 *Environ. Sci. Technol.* **322** 3379–3385
[8] Huang H L, Zhang S Z and Christie P 2011 *Environ. Int.* **159** 238-243
[9] Alkorta I and Garbisu C 2001 *Bioresour. Technol.* **79** 273-276
[10] Chaineau C H, Morel J L and Oudot J 2000 *J. Environ. Qual.* **29** 569-78
[11] Lu S J, Teng Y G, Wang J S, Sun Z J 2010 *Chemosphere.* **81** 645-50.
[12] Gerhardt K E, Huang X D, Glick B R and Greenberg B M 2009 *Plant Science.* **176** 20-30